Place Brand Formation and Local Identities

Staci M. Zavattaro

Routledge
Taylor & Francis Group

LONDON AND NEW YORK

First published 2020 by Routledge

2 Park Square, Milton Park, Abingdon, Oxon, OX14 4RN
52 Vanderbilt Avenue, New York, NY 10017

Routledge is an imprint of the Taylor & Francis Group, an informa business

First issued in paperback 2020

British Library Cataloguing-in-Publication Data
A catalogue record for this book is available from the British Library

Library of Congress Cataloging-in-Publication Data
Names: Zavattaro, Staci M., 1983– author.
Title: Place brand formation and local identities / Staci M. Zavattaro.
Description: Milton Park, Abingdon, Oxon ; New York, NY : Routledge,
    2020. | Includes bibliographical references and index.
Identifiers: LCCN 2019014901 | ISBN 9781138500044 (hardback) |
    ISBN 9781351013512 (ebook)
Subjects: LCSH: Neighborhoods. | Branding (Marketing) |
    Place attachment. | Group identity.
Classification: LCC HT153 .Z39 20120 | DDC 307.3/362—dc23
LC record available at https://lccn.loc.gov/2019014901

ISBN: 978-1-138-50004-4 (hbk)
ISBN: 978-0-367-78554-3 (pbk)

Typeset in Times New Roman
by Apex CoVantage, LLC

Place Brand Formation
and Local Identities

This innovative book explores micro-level neighborhood branding and the creation of distinct local identities in neighborhoods.

It begins by situating place branding literature at the neighborhood level and then gives consideration to what the core components of a neighborhood brand might be. It does so by drawing on extensive interviews with key actors in the United States, such as government officials, Realtors, economic development professionals, urban planners, and neighborhood residents. Core topics such as belonging and community, identity, nostalgia, idealism, and recreation are explored. The book concludes with a proposed working definition of neighborhood brands and branding that stakeholders can use to promote and market their neighborhoods accordingly – or avoid branding them entirely.

This book offers a novel contribution to place branding and destination management literatures by moving beyond the dominant macro-level narratives. It will be of interest to scholars and students studying in urban planning, tourism, destination branding, marketing, public administration and policy, and sociology.

Staci M. Zavattaro, PhD, is associate professor of public administration at the University of Central Florida. Her research areas include place branding, administrative theory, and social media use in government. She serves as editor-in-chief of *Administrative Theory & Praxis*.

To Dr. Patricia Patterson, for showing me a love
for qualitative inquiry.

To my mom and dad – try reading this one, will you?

Contents

Acknowledgments

I thank all the people who took time to speak to me for this research. Many people were quite generous with their stories, and I hope I did justice by sharing them here. All people are anonymized, including local officials who said I could name them, but I chose not to for consistency throughout. Any mistakes are my own.

Introduction

I think my roommate and I chose a better neighborhood because if we had chosen directly in the bar district, it is super loud. I don't have to be in that area if I don't want to. I think that was what was so important with my identity is having the ability to choose and not have it chosen for me.
– Victory Park, Dallas, Texas, resident

As I write this, I sit at my teal-blue desk looking out a window at Lake Baldwin in Orlando. Baldwin Park is my chosen neighborhood, redeveloped on the site of a former US Naval training facility. What was once an abandoned brownfield is now a thriving neighborhood known mostly for its million-dollar houses. When I tell people I live in Baldwin Park, I usually hear things such as "Oh wow" or "Well, that's nice." I feel defensive for some reason, reminding people I live in a one-bedroom apartment with rent comparable to other complexes in and around Orlando. The name Baldwin Park has its own brand images and associations that are projected onto me simply because this is where I chose to live.

Sometimes we take the things around us for granted. Our neighborhoods could be falling into that trap because they are sometimes so natural to us we might not think twice about them. We might see our neighborhood as a place where we bought a home and can sleep peacefully. We might see it as a temporary stop in an apartment complex because it is close to work. It is where we live because our home is walking distance to bars or restaurants. For parents, good schools are often crucial to picking a neighborhood. We are told all the time to live in a place that has a combination of live, work, and play options. In some cases, we pick a neighborhood because that was the only place we could find somewhere to live in a shrinking market.

I grew up in South Florida, lived in South Texas and Mississippi, then moved to Orlando in 2015. Each neighborhood has been different for many reasons. In Brownsville, Texas, I chose to live in faculty housing apartments on campus so I could easily get to and from my office. There really was no neighborhood in the traditional geographic sense or the community sense either. My neighborhood in Starkville, Mississippi, was the first time I encountered people waving as I drove along my street. Waving simply to say hello. In South Florida, that rarely happened. It was a bit unnerving at first because I was skeptical: What do you want? But it came to be a nice sight when I came home from a long day working on campus. Neighbors would talk to each other, help each other. I did not experience that where I grew up in South Florida.

For the start of my career, I described the neighborhoods I lived in as "fine." They were fine. My faculty housing apartment was great, but I really did not see many neighbors. I walked to my office and back. In Mississippi, the house I rented was quite old – and sometimes dangerous because of it. (The shed out back where I had to do laundry almost caught fire, but it was a neighbor who alerted another neighbor who texted a friend who called me to tell me of the impending danger because I was not at home. Really, that is the power of neighbors.) There were no sidewalks, which I found out the hard way when the sidewalk ended as I was walking to vote one year. I had to shuffle along in the grass and hope I did not step in a hole.

When I moved to Orlando, I told myself I was done with fine. I wanted somewhere safe, walkable, not too close to campus but not too far, and with restaurants nearby. Little did I realize that living in Orlando is relatively expensive, as renting a house in Mississippi's college town will warp your sense of cost of living quickly. The budget I set put me in apartments in areas that I did not feel safe walking around – my perception of safety. I upped the budget and found an apartment in Baldwin Park. What I like most is the 2.5-mile paved path around Lake Baldwin. I put on my running shoes and head out. No car needed. I can walk to the restaurants in the town center, and my kickboxing gym is in that retail core as well. The November Project Orlando workout group meets near a restaurant in the same town center area, so I walk there to participate in the workouts, too. Finally I was done with fine.

Orlando's neighborhoods give the city a sense of identity beyond the theme parks, which, by the way, you never would know are so close by if you do not make the drive down there. To give context, Baldwin Park is about a 40-minute drive away from Walt Disney World. Many visitors are surprised when I say I live that far away, then they ask, "Wait, I thought you lived in Orlando?" I do, but the city is large and spread out. The neighborhoods create that sense of place, for good or ill. Many people have taken serious steps to create that identity through the years, to embrace what was there or change what they did not like. I quite enjoy living in Baldwin Park because of the outdoor spaces, the walkability, and the friendly people. (I often see the same people on the walking path around the same time each morning when I walk to kickboxing or for a run.)

Looking at Orlando's neighborhoods inspired this research, as many neighborhoods have unique identities. Mills 50 is trendy and cool, with hip local restaurants and shops. Family businesses dot a main street, with some dating back to the 1950s and 1960s. The Lake Nona neighborhood is new, focusing on its medical city image because of proximity to hospitals and marketing its connected trail system to promote personal health. College Park is older, more established, with bungalow-style homes. Parramore is, unfortunately, known as the area with high crime, as is Pine Hills. For better or for worse, neighborhoods like cities, states, and nations have strong identities that can help or hinder their progress. Yet in the literature there are few studies that examine neighborhood branding, identity, and image.

Wherry (2011) goes in depth with one neighborhood in Philadelphia to begin answering questions about neighborhood branding. He reminds us that neighborhoods are indeed important to study given their proximity to us, cultural relevancy, and vibrancy. He explores the direct and indirect measures people in the Philadelphia barrio took to change and cultivate the neighborhood brand. My task here builds on that idea, as I expand the geographic reach from one neighborhood (Wherry, 2011) to neighborhoods throughout the US. The purpose of this book is to better understand the concept of neighborhood branding, identity, and image from the perspective of professionals and residents. I spoke to 75 people from across the country living and working in neighborhoods of various sizes and reaches. People came from coastal cities as well as Midwestern states. I tried to speak to as wide a variety as possible to understand how people relate – if at all – to their neighborhood brand identity.

A working definition of a neighborhood brand appears in the conclusion. Why? The book's chapters are an accumulation of what people said makes up a neighborhood and subsequently its brand. I outline each in detail before arriving at a working definition of neighborhood branding based on these factors. The chapters come from interview transcript content analysis and show the most commonly named elements of neighborhood identity. Interviewees included government officials, Realtors, economic development professionals, urban planners, and neighborhood residents. Subjects come from neighborhoods in large cities such as Dallas, Orlando, and New York, as well as smaller cities in Louisiana, Florida, Arkansas, and California.

I did not control for things such as neighborhood population or someone's age, as that was not the point of this study. This is an exploratory study to better understand and develop a working definition of neighborhood branding and identity. Therefore, I wanted to talk to as many people as possible from various geographic backgrounds, stopping when reaching a point of data saturation. I then transcribed and analyzed the data, leading to the patterns presented in this book. The aim here is not to generalize; instead, this is a chance to conduct an exploratory study using qualitative interviews given the topic's newness.

The book proceeds as follows. In Chapter 1, I give an overview of place branding literature, showing how that work expands into the neighborhood level. I also share information from NeigborWorks America, a nonprofit organization dedicated to neighborhood well-being, about their practical applications of neighborhood branding. As you read along, you will see why defining a neighborhood was difficult for the majority of interviewees for this research. Chapter 2 begins the components of a neighborhood brand. In Chapter 2, I examine sense of community and feelings of safety, two of the most significant patterns to emerge during data analysis. Sense of community was an interesting qualifier, given that *feelings* of belonging sometimes outweighed *actual* belonging. Safety, too, was often more in someone's mind rather than based on firm crime statistics. Taken together, they are emotional reasons someone might choose a neighborhood, and these emotional reasons can indeed be marketed and promoted to increase residency.

In Chapter 3, I explore the role of government-neighborhood relations, examining how cities and residents work together to create healthy and thriving neighborhoods. Given my expertise is in public administration and public branding, this chapter is important because it shows the connection among branding, neighborhoods, local governments, and nonprofits. There is a symbiotic relationship happening that is crucial to creating healthy communities and given the complexity, I found an interaction among residents, cities, and nonprofits. For many, this connection comes in the form of digital communication, but there is also a built environment component critical to forming meaningful neighborly connections. Community events, too, play a role here.

Chapter 4 focuses on the role of nostalgia and idealism in neighborhood branding. Both play a fanciful role in constructing neighborhood identity. Nostalgia, for many, is what they believe a neighborhood should be, while idealism is seen as a mechanism to keep people out (either in the form of actual deed restrictions, or neighborhood offerings such as retail stores that cater to a certain class). Gentrification and its relation to neighborhood branding are discussed in this chapter. Chapter 5 delves more fully into the role of parks, recreation, and green space related to the neighborhood brand. Many interviewees mentioned choosing a neighborhood for its easily accessible recreation features – even if they might never use them or use them rarely. Chapter 6 highlights what I called coded language related to neighborhood identity. This is the kind of language people used to separate themselves from others. Usually this language had racial undertones that indicated the "good" from the "bad" neighborhoods. I was not sure how to tackle this in the book, but the wording came up enough that I wanted to address it. This is important when it comes to branding because it shows the power of word-of-mouth and imagery when it comes to creating brand associations.

Finally, in Chapter 7, I offer a working definition of neighborhood brands based on a combination of the factors outlined in the chapters. The definition can be a starting point for others undertaking this kind of work. As branding studies grow in many fields, it is important to begin to work toward a starting point. I know the definition can be refined or refuted – and I welcome either. I also encourage you to engage with me about your neighborhood by following me on Twitter @StaciWithaZ. Share photos and stories of your neighborhood and let me know how else I can begin to understand the complex realities of neighborhood living.

1 Place branding and neighborhood identity

I think in terms of branding, and it is important for a neighborhood to have a brand; it just depends on the person. Newer families buying into newer neighborhoods, there's a shift in terms of how important is the brand or the name of where you live. When you talk about the brand of the neighborhood, it really is the name. It's not necessarily a logo. It's where do you tell people you live?
– Audubon Park, Orlando, resident

Sometimes identifying a public brand is relatively easy. When one thinks of Paris, they think of food, the Eiffel Tower, wine, and culture. New York City conjures hustle and bustle, business, multiculturalism, and, unfortunately, terrorism. Orlando is home to theme parks, Mickey Mouse, and fantasy. There are countless examples, but the bottom line is that these few cities named have a strong brand identity – for better or for worse. While most studies examine nation, state, and city branding, I look at the micro level of the neighborhood to better understand how neighborhood brands form and what they do for those areas. This chapter outlines definitions of a neighborhood, provides an overview of place branding and marketing, and then briefly explains neighborhood branding and identity related to place attachment.

What is a neighborhood?

No question I asked people gave them more pause than "What is a neighborhood?" I did not expect this reaction, but upon examining the literature it became clear there is no set definition of a neighborhood. People in this study usually defined it in terms of a geographic location, a sense of community, or both. Some examples of definitions from the interviewees include:

Hmm, that's a tough question. A neighborhood is obviously a place where people live so they sleep there at night. But a neighborhood is also a community of people who can [band] together and help people out when they need it.
– Urban planner and Audubon Park, Orlando, resident

So a neighborhood is numerous homes clustered together with a sense of friendliness and collegiality, camaraderie among the people who live there. To me a neighborhood is safe and comfortable and serene.

– Celebration, Florida, resident

For me your neighborhood is your immediate walking area. I can go to other neighborhoods, but my neighborhood would be what I can get to in a few minutes. The adjoining residential streets next to me that would be part of my neighborhood. The things within half-mile walking distance, that would be my neighborhood.

– Dallas, Texas, resident

A neighborhood is an encompassing community that brings the environment, people, social, intellectual, together, hopefully dependent upon the community for the good of those who live there. There's a level of cooperation, collaboration, reliant upon what people have to contribute.

– Baldwin Park, Orlando, resident

The physical and limited place that we see. That's the first idea of what we think of a neighborhood – a name and geographic location. It's similar to the idea of what is country? Probably the first thing we think is where it's located. Following the parallelism with the country, I think that is has to have economic activities that define it, a very specific economy or lack of economy. Economic activities define what is a neighborhood, and it defines the types of people.

– Glen Elm, Tucson, Arizona, resident

Boy there's a question. A lot of it's just geography, its identity within that geography. In my experience, neighborhoods should have a central point where everybody knows to converge during certain times of the year.

– Main Street director, College Park, Orlando, Florida

That's interesting. That could be a somewhat amorphous concept. I would think a local place identity really tied in with a feeling of ownership and whether that is an identity of 'This is where I live,' or 'This is my community,' or how this particular area acts or feels.

– Economic development expert, Washington, DC

That's probably the million dollar question right there because that can shift. Different people can live in the same neighborhood and define it differently. I think often the definition is sometimes arbitrary. Sometimes it comes from outside of the community. Sometimes it's defined by some sort of government process. If there are no geographic markers, it's hard to say someone who lives on the other side of the block, are they part of the neighborhood or not?

– Community organizing expert, Washington, DC

That is a good question. What is a neighborhood? I would say a neighbor-hood is some place where you have a clear identity that's either established by history, geography, personalities, common interest, diversity or all of the above. I see it as some kind of unifying characteristic where you can have a unique identity.

— Millers Bay, Oshkosh, Wisconsin, resident

To study neighborhoods, the US Census Bureau uses tracts, "small, relatively permanent statistical subdivisions of a county or equivalent entity that are updated by local participants prior to each decennial census" (U.S. Census Bureau, 2012, p. 1). Many dictionary definitions of neighborhoods also exist, but one from *National Geographic* magazine stands out (2018, pp. 1–2):

A neighborhood is an area where people live and interact with one another. Neighborhoods tend to have their own identity, or 'feel' based on the people who live there and the places nearby. . . . Neighborhoods often have fuzzy geographical boundaries, so sometimes it's difficult to tell where one starts and another ends. Major streets often act as logical boundaries, but people usually define a neighborhood by its characteristics.

The definition aligns with many interviewees' perceptions of a neighborhood – geographically bound though potentially hard to tell, with a strong identity, and a sense of connection. Neighborhood definitions usually fluctuate depending upon how planners or scholars need to view the places (Kallus and Law-Yone, 2000). As Kallus and Law-Yone (2000) explain, the neighborhood as a conscious architectural unit emerged with the introduction of modern urban planning techniques, so neighborhoods are always part of a larger urban planning and design program. Neighborhoods by nature are smaller than the cities in which they are developed, so it is through neighborhoods where connections can form. "Accordingly, the neighbourhood becomes a convenient and easily defined urban area; a clearly bounded territory, a cluster of streets, and a service area, which generates a social network" (Kallus and Law-Yone, 2000, p. 817). Neighborhoods express the tension between professional planning and management, which attempts to control the urban environment, and social connection and cohesion, which is more independent and voluntary. Neighborhoods, then, become a "marketable product" (p. 818) Realtors and cities can sell as an economic success story.

Choguill (2008, p. 42) argues the definition of a neighborhood is "in the eyes of the beholder" because it can involve geographic boundaries, class, religion, race, or myriad other factors that could define the space rather than only set physical features. Similar to Kallus and Law-Yone (2000), Choguill (2008) also traces neighborhood development to modern urban planning, specifically to the emergence of garden cities in Britain as a means to counter crowded central cities. Thanks to Ebenezer Howard, the movement's founder, garden cities would be their own contained units with housing and economic opportunities, along with chances for agricultural activities in a green belt surrounding the community to keep it safe (Choguill, 2008). Howard's ideas transformed urban planning into a focus on spatial arrangements rather than public health as a foil to crowded cities. As the concept developed and was refined in Britain, America, and elsewhere, the focus turned to neighborhoods as spaces for connection rather than solely as dwelling units.

Melvin (1985) argues that the American neighborhood, however, changed the culture of cities from spaces of walkability and connection to the seeming enemy that required suburbanization. Walkable cities were the epicenter of social, religious, political, and economic thought and practice. The introduction of automobiles changed that, allowing people to live farther away from a city center in a suburban neighborhood. Immigration and a division of socioeconomic status further ruptured walking cities (Melvin, 1985). Neighborhoods fell in and out of favor with a move toward metropolitan cities in the 1930s in the US. Then again the new urbanism movement with a focus on walkability and density also further pushed aside traditional suburban neighborhoods. She illustrates how neighborhood identification and study changed throughout the country's development, along with swings in urban planning favor. All told, there is no agreed-upon definition of neighborhood, and the definition depends upon the era, how someone sees the neighborhood, and what someone wants to understand about the place.

Mumford (1954), in a classic treatise on neighborhoods, argues that "in a rudimentary form neighborhoods exist, as a fact of nature, whether or not we recognize them or provide for their particular functions. For neighbors are simply people who live near one another. To share the same place is perhaps the most primitive of social bonds, and to be within view of one's neighbors is the simplest form of association" (p. 257). Similar to Melvin (1985), Mumford (1954) traces global historical developments in neighborhood planning, noting some experts wanted neighborhoods to center on a physical space, like a school, while others focused more on the social aspects of neighborliness. Neighborhoods, he argues, are boons to central city development because they take the stress off these cores, which could become (and did become) crowded and unsafe. In sum, Mumford argues that neighborhoods develop whether purposefully planned or not, so scholars and practitioners can try to "solve" neighborhood problems all they want but the unit will exist in some shape or form.

Place branding, image, and identity

Despite no single definition, neighborhoods are distinct places. Fields such as sociology, urban planning, environmental geography, medicine, and more use neighborhoods as units of analysis. As a public administration scholar, I am interested in the governance and engagement aspects of neighborhoods. I want to know not only why people choose places to live, but also how neighborhood entities and local governments work to strategically form neighborhood identities. I want to know the role that nonprofit organizations and other stakeholder groups (residents, business owners, etc.) play in this process. While the book is about neighborhood branding and identity, it also is a study in administrative practices at the micro level of the neighborhood. Given the core focus is on neighborhood branding and identity, I offer here a brief overview of place branding and identity to show where the research fits.

Specifically within public administration, place branding is becoming a key governance strategy that relies on networking, co-production, and shared governance for success (Eshuis and Klijn, 2012). Within a political realm, people might be more accustomed to seeing branding efforts of political parties or candidates, but within local governments more study is needed (Eshuis and Klijn, 2012). First, the terms *brand* and *branding* are complex and find natural homes in the corporate world. A branded product, company, or service has some kind of identifying name or trademark. Think about Apple as a corporate brand, along with its iPhone as another distinct product brand. If people love the overall Apple brand, they are likely to buy its various branded product lines (iPad, Mac book laptop, etc.).

Branding refers to the active process of developing and communicating the brand attributes. Anholt (2005) notes there are three ways to understand branding: popular, simple, and advanced. The popular way is as a marketing umbrella term referring to selling and promoting items. The simple way refers to logos, images, or slogans to identify a brand. And the advanced way takes more behavioral and emotional aspects into account, using branding strategies to build and cultivate relationships with consumers (Anholt, 2005).

Place branding ideally takes the principles of the latter strategy and applies them to nations, states, country products, tourism destinations, neighborhoods, and more. A challenge is that within the public sector, there is an aversion to branding because most people adopt the simplistic definition of only promotion and selling (Anholt, 2005). Despite nomenclature challenges, place branding as an activity is becoming increasingly necessary in a globalized world that thrives on competition for often-scarce resources (Hanna and Rowley, 2008). Kotler, Haider, and Rein (1993) explain that public sector marketing and branding are a natural extension of business-based logic, which seeped into the public sector fully in the 1990s (Osborne and Gaebler, 1993). Within public administration, branding and marketing also are outgrowths not only of globalization but also movements such as New Public Management and Total Quality Management that advocated more business-like practices in government (Eshuis and Klijn, 2012).

A *place brand* is an emotional connection that serves as a cognitive shortcut to help someone identify the place in question, be it a country, destination, or neighborhood (Anholt, 2010). Place branding is the process of doing something to shape and communicate a brand identity, which often includes visual elements such as a logo and/or slogan (Anholt, 2010). Anholt (2010), though, is clear to say a place brand, just like a product brand, is not only a logo or slogan; those are helpful visual identifiers, but a place brand is more about feelings and connection. Related, the *brand identity* is what the organization attempts to communicate about its brand, while *brand image* refers to how people interpret those messages (Anholt, 2010). The success of a place or even product brand depends largely if not wholly on the people making purchasing decisions.

The public sphere is already complex, given the number of stakeholders who are invested in their cities, states, and countries. Related to place branding, interested stakeholder groups could include the city (or governing entity), nonprofit organizations, residents, business owners, the tourist board, hotels, museums, parks, restaurants, shops, etc. (Eshuis, Braun, and Klijn, 2013). All actors involved will usually have differing approaches to success, thus making any governance activities tricky and time consuming. Political and popular support, or lack thereof, is another challenge related to place branding as a governance strategy (Eshuis et al., 2013). For example, lack of political support might mean projects are underfunded, not funded, or removed entirely from the political agenda. Finally, another challenge, similar to what Anholt (2005) articulated, is that public agencies often are wary of the terms *marketing* and *branding* because they seem like mere rhetorical tricks rather than strategic processes (Eshuis et al., 2013).

Public administrators, then, might shy away from place branding efforts for fear of negative backlash. As an example, the city of Pittsburgh in 2019 rebuffed place branding efforts, calling them "a total waste of money" given the six-figure price tag for professional branding assistance (Smeltz and Murray, 2019, p. 1). The branding proposal meant to streamline visual identities for the city into a unified whole, but council members balked at paying $100,000 for outside branding services in a potentially trying fiscal climate (Smeltz and Murray, 2019).

Stakeholder involvement in public branding projects often is more vital than when compared with private sector counterparts. Government accountability requires openness and transparency, especially concerning fiscal matters. Within the public sphere, branding efforts usually are meant to attract residents, tourists, business owners, corporations, and other investors (Klijn, Eshuis, and Braun, 2012), so interaction between myriad stakeholders is not a necessary evil but a mandated action. Public administration scholars usually call these collaborative actions collaborative governance or co-production, whereby actors work together ideally with equal power to arrive at solutions to complex problems (Agranoff and McGuire, 2001). Klijn et al. (2012) import this same logic into the place branding process, as "those actors often have different perceptions about the place to be branded, the desirable solutions to problems and different ideas about the desirable image to be communicated" (p. 505). Klijn et al. (2012) find that stakeholder involvement improves brand performance and the ability to attract different target groups. Just as in public administration processes, stakeholder involvement in branding processes matters.

A simple Google News search reveals the prevalence of place branding and marketing strategies at all levels of government. The city of Painesville, Ohio, recently released its new branding campaign as a way to counter negative images about the city. Using an outside consulting firm, the city launched in 2019 a new brand logo and associated imagery. According to a press release, "This is an exciting time for our city and the new branding really reflects the changes we are making," says City Manager Monica Irelan Dupee. " 'Our new brand and communication efforts will help our residents, business owners and stakeholders relay the stories that celebrate our city's qualities, which will then attract others to participate, contribute and invest in what makes our city unique'" (City of Paines-ville, 2019, p. 9). The quote is emblematic of an economic-based view of branding success, focusing on investment within the city.

Similar to the case in Pittsburgh, some elected officials in Spring Hill, Tennes-see, objected to a branding measure that cost $66,000 for a private firm, originally budgeted at $50,000 (Willis, 2019). When describing the efforts, the story writer put "place branding" in air quotes, as if to indicate this is not a real phenomenon. The story details tensions among elected officials when it comes to the city's branding efforts, noting the cost and lack of measurable outcomes of success. Return on investment for place branding efforts always is tricky (Zenker, 2011), so it is no surprise officials would express concern.

Neighborhood branding and identity

The foregoing few examples show both the pervasiveness and trickiness of these practices at the local level. While cities are interesting to study, I drill down fur-ther into the neighborhood level, as these are places where people spend most of their time and make one of their biggest investments. Not surprisingly, there is no agreed-upon definition of neighborhood branding either, likely because of the conceptual fuzziness surrounding the terms *neighborhood* and *branding* as separate entities.

As detailed previously, branding is an active governance strategy to develop and communicate a place identity (Anholt, 2005). I spoke with a representative from NeigborWorks America to learn more about how this nonprofit organization understands neighborhood branding. In their documentation explaining neighbor-hood marketing and branding, NeighborWorks America define a neighborhood brand simply as "what people think of your neighborhood" (Kelsh, 2015, p. 7). NeighborWorks America launched its neighborhood branding and marketing pro-gram in 2012 after the housing recession in the US. The organization relies on partners to work with local communities on their targeted needs, explained the executive to me. In the aftermath of the foreclosure crisis, neighborhoods and cities were dealing with blight, abandonment, and disinvestment. Even though some areas received grant funding, they were "still struggling to attract homebuy-ers because of the perception" so the idea to help neighborhoods with branding and marketing was born. "When we train, we coach that all neighborhoods have a brand already. It exists," he said.

"It is part of our ethos, our DNA, that the community development organiza-tions that comprise our network came out of resident-led initiatives," he said. "We care about the residents in these places. Our starting point is the residents." The first step for NeighborWorks America representatives is to establish a local neigh-borhood branding team made up of concerned and interested stakeholders, which could include people such as home owners and renters, business owners, Realtors, elected officials, and those with media savvy, the NeighborWorks America repre-sentative said. "That group goes through a process of understanding the current brand. If that's not favorable, then what are some concrete steps to take? They (the partners) give organizations training on all of this and a marketing consultant who works with organizations to coach them through."

He said community building is the core of their efforts at NeighborWorks America so investing in these long-term strategic processes is worth the time, effort, and investment up front. "My position is that it's worth that time because there's risk of the whole effort going off the tracks if you haven't consulted those key stakeholders in the beginning." This view aligns with existing research into place branding best practices (Kavaratzis and Hatch, 2013). The NeighborWorks America process also involves extensive market research, communication, and evaluation of neighborhood branding efforts. Team leaders are left with tools and resources to continue the efforts even after NeigborWorks America officials have left. "We are most successful in places where these efforts result in really robust collaborations. The bigger result is that organizations came together around a shared purpose and worked collaboratively. That's [what's] going to have a bigger long-term effect."

He advises people to start with the neighborhood image because "oftentimes that is overlooked. The image drives behavior. . . . Images drive what people invest in you, so where I start first is understand what your current image is." Similarly, Farris and Kendrick (2011) argue that brands are a collection of stories people tell and weave together to create an image of the place or product. For Farris and Kendrick (2011, p. 6), a neighborhood brand is defined as "recognition among residents that a particular business excels delivering a product, selection, service, or experience" (p. 6), while acknowledging that neighborhood brands are socially constructed as a result of the stories and perceptions. This is the same point the NeighborWorks America representative was making regarding the importance of external images and perception.

While NeighborWorks offers a definition of neighborhood branding, a handful of academic studies are examining the topic. Wherry (2011) captures locally driven branding efforts in a Philadelphia barrio called Centro de Oro. Using an ethnographic approach, Wherry (2011) embedded himself in the community and interviewed stakeholders he called cultural entrepreneurs who took active steps toward improving the neighborhood's physical offerings and associated negative images. His findings showcase the complexity and connectivity inherent in many place branding efforts, echoing Eshuis et al's. (2013) writing about place branding as a governance strategy that necessarily involves myriad stakeholders with different outcome measures in mind.

Wherry (2011, p. 4) defined a neighborhood brand as "apparent from what businesses sell, how their storefronts are designed, what kind of music emanates from open neighborhood widows and passing cars, and what kinds of themes are depicted on the neighborhood's plentiful murals." The brand is an amalgamation of different aspects of neighborhood identity. The barrio was primarily Latino residents who often were stigmatized because of a low socio-economic status. The ethnic neighborhood, then, had a challenge of rebranding itself while also remaining true to its cultural roots. Wherry (2011) explains the inherent tension between the economic push for rebranding success and the historical imperative for authenticity among those with longevity in the neighborhood. So-called good rebranding, then, can potentially gentrify the neighborhood and push out historical and cultural aspects that made it unique in the first place (Wherry, 2011).

Similar to Wherry (2011), Johansson and Cornebise (2010) also studied an ethnic neighborhood, this time looking at Andersonville in Chicago. Andersonville, though not an officially designated city neighborhood, is a predominantly Swedish enclave that is slowly losing that identity because of out-migration and neighborhood flux. While many Swedes are leaving the neighborhood, Johansson and Cornebise (2010) argue that difficulty remains when trying to find a brand identity for the area given a desire to hold onto the community's characteristics. Taking matters into their own hands, local residents banded together to create events celebrating Swedish culture. Events were open to the public to ideally educate others about the neighborhood and culture. Despite these positive steps, neighborhood attrition remains, leading to the exploration of other branding strategies to spur economic development, potentially pushing aside the Swedish identity totally.

In another study, Masuda and Bookman (2018) note that neighborhood branding strategies are usually a governmentally driven endeavor to increase urban renewal and/or economic investment. They define neighborhood branding as "the symbolic and material practices of state and/or private cultural producers who aim to enhance the appeal of local areas within the city in order to attract investment, promote consumption, reduce criminality, or to achieve social and cultural aims such as invoking civic pride" (p. 166). The definition hones in on economic outcomes as measures of success, which is only one element of the branding picture (Zenker and Martin, 2011), especially when it relates to public administration (Eshuis and Klijn, 2012).

Neighborhood branding often is more complex than even city or nation branding given the hyper-local focus (Masuda and Bookman, 2018). People might focus more on perceived negatives, especially exacerbating flaws such as broken sidewalks or potholes. One member of the City of Orlando's Communications and Neighborhood Relations (CNR) team told me that she often cautions residents about posting negatives on social media because those posts can often go viral. "Is that the image that you want of your neighborhood? This is what all of Central Florida, a nine-county region, is going to hear about Orlando neighborhoods." At the same time, though, she explained that residents are important for maintaining neighborhood brand identity because they are the ones out there walking the sidewalks or parks and noticing something broken or amiss. City officials might not know because they cannot have a presence everywhere, so neighbors are key.

Building a community was at the center of rebranding efforts for Abbot Kinney Boulevard in Venice, California (Deener, 2007). Deener (2007) examines this popular street as neighborhood demographics shifted. He explains the road, lined with local businesses and retailers, became a dividing line, both physically and psychologically, for the neighborhood identity. As such, the place was often contested because of political, social, and cultural pressures (Deener, 2007). According to his interviews, the boulevard remains a racial dividing line while also serving as a source of nostalgia for longer-term residents. Newer residents, though, view the street as an opportunity to rebrand and reframe the neighborhood away from negatives images (Deener, 2007), marketing it as "the newest street in Los Angeles" (p. 298). To put physical changes in place, a neighborhood association formed, and membership fees went toward street beautification projects, including benches and bicycle racks. The association tried to use a bohemian vibe to promote the street and the neighborhood, but the gang violence nearby was sometimes too much to overcome and investors often shied away. "Newcomers make efforts to preserve the street's commercial culture, believing the anticorporate branding is central to Venice's bohemian identity, but they do not make equal efforts to protect the use of the street for long-time residents who cannot afford to shop at the new upscale commerce" (Deener, 2007, p. 293). Gentrification, then, was a real problem regarding the rebranding efforts for the neighborhood. This is not uncommon (Wherry, 2011).

Though not expressly about a neighborhood brand, Pais, Batson, and Monnat (2014) study the effects of neighborhood reputation after the housing crisis in the US, specifically what happened in the city of Las Vegas, which was hit hard by the foreclosure bubble. Positive neighborhood reputations, they argue, are vital for building overall strong cities, while the inverse relationship also exists. Similar to brand image, brand reputation depends upon residents' perceptions of the place (Pais et al., 2014). According to their findings, a positive neighborhood reputation played a mitigating role on the effects of the crisis. If a neighborhood already had a strong reputation, it was perceived as resilient after the foreclosures hit (Pais et al., 2014). There is evidence that a strong neighborhood reputation has important consequences for residents and the city as a whole.

As the news stories referenced previously show, place branding efforts (including those for neighborhoods) are difficult because the return on investment is often difficult to see. Taxpayers know, for instance, whether the city is repairing a sidewalk or building a new park. Those are physical enhancements easily recognized. Economic development, though, is often more difficult if not accompanied by physical buildings such as a new hotel. Rich and Tsistos (2016) found this occurred when looking at an arts district in Baltimore, Maryland, called Station North Arts and Entertainment District. Using stakeholder interviews, they found that many people perceived the rebranding efforts as a trick with little to no direct benefit on the neighborhood or city. Despite this, some interviewees did buy into the positive aspects of the revitalization, noting positive externalities and spillover from being near by a hip arts district (Rich and Tsistos, 2016).

Practically, neighborhood branding is not easy. A neighborhood in Belfast, Ireland, is feeling this strain as a developer wants to revitalize an area by calling it Tribeca. Now, Tribeca is a name associated with a geographic region in New York City – Triangle Below Canal (TriBeCa). In Belfast, the project would cost about $640 million in an area called Cathedral Quarter (Glassman, 2019). "I'm sorry but this is Belfast, not New York," the city's mayor, Deirdre Hargey, responded in a tweet. "We are proud of our city, its people, its place names and its heritage, that's what gives Belfast its soul!" (Glassman, 2019, p. 3). According to the article, the developers said their Tribeca stands for Triangle Beside Cathedral, but elected officials in Belfast were not buying it. Counter movements Save CQ have sprung up online, with residents opposing the rebranding (Glassman, 2019). Counter-branding is evident when people oppose having a brand given to them rather than being part of deliberative processes (Braun, Kavaratzis, and Zenker, 2013).

In another example, a neighborhood in Minneapolis wants to rebrand to give people a clearer picture of what – and where – it is. Nicollet Island-East Bank Neighborhood is wedged between two other neighborhoods and often gets conflated with them (Yoo, 2018). Some residents want to change that, suggesting calling the area Old Town – though unofficially at first. Without changing it on paper from Nicollet Island-East Bank, he (Scott Parkin, a local Realtor) said he's just trying to get Old Town to catch on, conversationally. "Word of mouth, grass-roots," Parkin explained. "Just trying to get people excited. Once they say it a few times, it's just natural" (Yoo, 2018, pp. 9–10).

One of the interviewees for this research described working with Neighbor-Works America to help her neighborhood craft an identity. She lives in the Millers Bay neighborhood in Oshkosh, Wisconsin. The neighborhood is on Lake Winnebago and serves as a central meeting point and recreational feature for the neighborhood. She explained the NeighborWorks America chapter eventually evolved into what is now called Greater Oshkosh Neighborhoods, Inc., but at the time the chapter helped Millers Bay create a new logo. A local company stepped in to help, but she said many neighbors did not like the proposed logo. "So I was the one who proposed the new one we have, and they tweaked that." The logo reads Millers Bay but a sailboat replaces the letter A in Bay. The boat hearkens, of course, to the nearby lake and lifestyle it provides. "We wanted to promote the waterfront. That is really what our neighborhood is all about. It really makes us unique." Now there are banners and neighborhood identification signs with the logo, along with landscape features such as walls or planters that also show the logo.

Place attachment

Place attachment, along with branding and identity, helps explain why people take such pride in their places – or not. Place attachment is "an affective bond or link between people and specific places" (Hidalgo and Hernandez, 2001, p. 274). The keyword there is "affective," meaning an emotional connection. Like brands, place attachment also serves as a cognitive shortcut for decision making. Place attachment manifests in neighborhoods because, according to Hidalgo and Hernandez (2001) people often show more signs of attachment to the micro level rather than a city or state. The concept is intricate, involving connections among physical space, emotions, and personal identity (Ujang, 2012).

Similar to what Pais et al. (2014) found regarding neighborhood reputation, place attachment also can have positive or negative mitigating effects (Comstock et al., 2010). Effecting neighborhood attachment are factors including, but not limited to, length of time someone resides in the neighborhood, owners versus renters, physical conditions of the neighborhood (broken windows, landscaping, housing style, etc.), community benefits (events, public pool, community garden, walkability, etc.), race, ethnicity, and more (Comstock, 2010). Drilling down to the block level, we see this relationship exacerbated because people often feel connected to their more immediate neighbors (Brown et al., 2003). Neighborhood attachment is usually higher the longer someone has lived somewhere (Bonaiuto et al., 1999) so that is why changes are so difficult because those alterations are often seen as an assault on a person's character and identity (Wherry, 2011).

Formal and informal associations within the neighborhood also are a key part of forming attachment (Woolever, 1992). Formal involvement includes activities such as serving on boards or committees, while informal activities can include simply chatting with neighbors or waving while out on a walk. Interestingly, Woolever (1992) found that density, a main component of new urbanism, reduced feelings of neighborhood attachment. Similarly, Talen (1999) found that new urbanism designs themselves do not necessarily foster a sense of community and connection but rather could attract the same type of people anyway, those looking for community features rather than an express connection. Interaction does not necessarily lead to place attachment (Talen, 1999).

Place attachment is an emotional connection that takes time to develop. It can be won or lost if people feel the place is no longer serving their needs. Neighborhood brand and reputation is indeed linked to neighborhood attachment, with people indicating they will move if a reputation weakens (Permentier, van Ham, and Bolt, 2009). Physical neighborhood change also needs to be coupled with emotional and even rhetorical change. There is a confluence and relationship between these factors. Only changing a logo or slogan is not enough to create a positive brand or reputation if there is something structurally flawed (Anholt, 2010). Therefore, neighborhood branding and identity are important to study as a governance issue because weak neighborhoods mean potentially weak cities.

Concluding remarks

This chapter briefly introduced the concepts of brand, branding, place branding, image, and identity. They all work together to create a mental image of brand attachment and loyalty. As the book moves along, I delve further into the patterns found while interviewing 75 people throughout the US about their neighborhoods. The patterns are elements that make up an overall neighborhood brand, which I define at the end of this research.

I asked people to explain the connection between themselves and their neighborhoods. The following is a sample of why people said they liked, or disliked, their neighborhoods based on a question of identity:

> *In some ways yes and in some ways no. It matches my identity in things that I like. It has access to the woods, to the highways. I can get to my job quickly. Things like that are all really nice. However, my politics don't match up with my neighborhood at all – at all. I mean like really, really vast differences. And that's the one thing I do find awkward is I'll be walking my dog and somebody will go on a political tirade, and I'm like whoops, keep my mouth shut.*
> *– Mt. Airy neighborhood resident, Philadelphia*

In some ways we all search for our identity at some point in our lives, or multiple points in our lives. I guess in some ways identity is a perception that others have of you that you may or may not agree with. I think that's certainly, in my experience, as I think of the neighborhoods in our area, some that have names just based on history or positive experience of the people who were living there.

– Economic development expert, Lafayette, Louisiana

I think that all neighborhoods and societies, they have their roots, that's where everything comes from. Local government specifically in the US should catch those values that might resonate with more people, so I don't think it's possible for a local official to shape an identity that doesn't exist from any-thing. They can echo voices or aggregate voices of the area.

– Glen Elm neighborhood resident, Tucson, Arizona

I think that living in an area that is so vibrant with so much stuff to do again within walking distance, it's easy to say, 'Oh, I live in the middle of nowhere so I don't do a lot of stuff,' but for us we live in a cultural hot pocket where we can do all this stuff like any day. I think my personality does fit this area because I like to do a lot of stuff.

– Baltimore, Maryland, resident

These are a few quotes to show the relationship between the self and neigh-borhood. Public administrators can learn from this by deeply exploring personal connections to place, asking people for their views in a co-productive manner rather than sometimes exploring top-down options. Neighborhood branding is an emerging area of study in public administration, so this research helps bring that gap into focus a bit while also leaving us with questions for future research.

2 A sense of community and feelings of safety

> *I think in terms of a safe neighborhood, I think it's definitely one that has a sense of community and has neighbors that look out for each other. It takes a village to raise a kid, and it takes a village to keep your neighborhood safe.*
>
> — Planning professional, Orlando, Florida

The two most prominent patterns in the data related to neighborhood branding and identity included what I coded as a sense of community, and feelings of safety. In this chapter, I explain each in more detail, showing how they often work together to create neighborhood attachment – and elements that can be packaged and sold through marketing and branding tactics.

Sense of community – A complex interaction

The first major pattern revealed during data analysis was coded as a sense of community. This is not surprising, given neighborhoods, ideally, could create a sense of belonging through urban design and collectivism. Jane Jacobs (1961) famously shared this ideal in her study of Manhattan neighborhoods, arguing that neighborhoods need fluidity built in. There must be opportunities for people to interact, to look out for one another – to maintain and invest in their place. Neighborhoods must be diverse not only in people but also in opportunities so people will want to stay even when life circumstances change (such as a new job). "A city's collection of opportunities of all kinds, and the fluidity with which these opportunities and choices can be used, is an asset – not a deterrent – for encouraging city-neighborhood stability" (Jacobs, 1961, p. 139).

This collection of assets and the ability to use them makes for a vibrant neighborhood for people who live there and beyond. Interviewees expressed this mostly through wanting a sense of community. People either mentioned the term explicitly or implicitly. Sense of community is tricky to define and sometimes even trickier to implement. In American culture, sense of community seems almost a paradox given our inherent values tend toward individualism rather than collectivism. Jacobs (1961, p. 117) articulates this conundrum well:

Let us assume (as is often the case) that city neighbors have nothing more fundamental in common with each other than that they share a fragment of geography. Even so, if they fail at managing that fragment decently, the fragment will fail. There exists no inconceivably energetic and all-wise 'They' to take over and substitute for localized self-management. Neighborhoods in cities need not supply for their people an artificial town or village life, and to aim at this is both silly and destructive. But neighborhoods in cities do need to supply some means for civilized self-government. This is the problem.

Being in a neighborhood though, for some, means wanting to belong, wanting to feel engaged. Indeed, Jacobs (1961) would argue this engagement is critical to neighborhood safety and survival. Creating vibrant neighborhoods is an interesting dance between locals and government officials who often control policy implementation, taxes, and approvals of neighborhood happenings and wants. Although these commitments can be costly for local governments, a strong sense of neighborhood community has the potential to encourage citizen participation, reduce citizen attrition, and promote economic development. Importantly, many of the government officials I talked to discussed the complications associated with intentional community development versus communities that develop organically.

I did not ask people directly about a sense of community. Instead, people would bring it up so I would probe for additional information. Some quotes represent varied views regarding a sense of community:

When you see people out walking around it's, 'Hi, how are you?' They're taking care of their yards, so there's pride there. They feel comfortable going next door to borrow an egg. . . . I think some of the amenities that contribute to that sense of community are gathering places, like a park that has a right of way, places to go grab a cup of coffee and see neighbors and be able to walk there or get in your car and you're there in two minutes. I think when you see an active neighborhood where there are people out walking, jogging, and kids out playing that really adds to that sense of community. It feels like, wow I stepped back in time.

– Member, city of Orlando CNR team

Togetherness, several local businesses there for community conversations or community issues. A hub for local officials or even like down-home local cuisine or very much focused on the local sports team. It's not just like a place like a Starbucks where you come in and get coffee or you're sitting at that chain restaurant. It's things that you want to support frequently. It's places where you can sit down and say, 'Hey, that's my neighbor.'

– East Riverside, Austin, Texas, resident

What I feel right now honestly is most of my neighbors are different. They're barely here. They don't feel a deep sense of connection to the community so they sort of keep to themselves. Two houses on either side [of us] converted into condos. When you live in a condo, your relationship to your front yard is really different. My wife spends a lot of time in the front and back yards. She has gotten increasingly upset over the last few years because our neighbors just don't give a shit about their yard anymore. They're not going to be out there in the same way that we are.

– Community organizing expert, Washington, DC

Human beings are sort of driven to coexist peacefully with their neighbors. They have the same concerns or goals for wherever they live. So it's where do we link arms collectively to try to improve the space that we all share? Do we know each other? Do we share common concerns in the first place?

– Main Street director, Orlando, Florida

I think for me, sense of community I just think about raising three kids in the city. It takes a village. It's people having your back. With Nextdoor, first of all, I think it's cool because I can connect with people with similar interests. Because they're my neighbors I trust those recommendations. I use it like three times a day for local recommendations, and I've never been let down by that. Community is about being safe, being stronger, being more connected – and all of those things have happened for me as a result of Nextdoor.

– Nextdoor executive

A sense of community is to have at the very base a sense of shared setting. And there are so many more ways to build community. It is incorporating where you live as a part of your identity, and if you don't want to do that then you move.

– San Diego, California, resident

Sense of community and neighborhood attachment are often linked (McMillan and Chavis, 1986), but the exact form of neighborhood design to aid in those connections remains unclear (Talen, 1999). Neighborhood design can influence acts of neighboring, like helping each other out or sharing a friendly conversation, as well as neighborhood attachment and meaningful social ties (Rogers and Sukolratanametee, 2009). The design effects, though, are mitigated when demographic variables are considered so new urbanism-style planning is not a panacea to create neighborhood sense of community (Rogers and Sukolratanametee, 2009; Talen, 1999). Like all things, context matters in neighborhood design and planning (Yang, 2008).

Neighborhood sense of community

While there are many ways to examine sense of community, for this research I focus on neighborhood sense of community. Perkins and Long (2002) explain that sense of community is an individual-level construct that works in concert with a broader social capital framework. A central component of neighborhood sense of community is trust in your neighbors, which could lead to increased collective efficacy and more neighboring behaviors (Perkins and Long, 2002). Foundational to a neighborhood sense of community is place attachment, as it creates bonds, motivates political activity, and increases economic and capital investment in a community (Perkins and Long, 2002). Oftentimes, interacting regularly with one's neighbors makes them more confident in the place and likely to stay longer term (Varady, 1986).

Informal neighboring activities, such as chatting while out walking or block parties, often were antecedents to creating more successful formal neighborhood organizations and partnerships (Unger and Wandersman, 1983). This relationship, though, could be tricky as Unger and Wandersman (1983) found that neighbors with strong, long-lasting informal ties are less likely to seek a formal organizing mechanism given they already feel invested and attached to the neighborhood. Less active neighbors, though, were found to have potentially meaningful organizing and managing skills that would lead them to seek formalized involvement paths rather than informal ones (Unger and Wandersman, 1983). Similarly, these informal networks of neighbors often contribute to a strong network of associations and neighborhood resources, especially for those considered outside the power majority in the neighborhood or region (Smith and Smith, 1978). Sense of community is found across different types of neighborhoods, including rural areas that buck traditional high-density new urbanism designs (Robinson and Wilkinson, 1995).

In planning terms, the neighborhood serves as a physical geographic boundary and space that often delineates how services are provided, yet Tannenbaum (1948) advocates for viewing the neighborhood as an equally important social space. Neighborhoods, when viewed through a social lens, can serve as a foil to loneliness and isolation. Germane to the research shared in this book, Tannenbaum (1948) also explains the importance of neighborhood identification: "The neighborhood unit, easily identifiable and standing out in relief against the rest of the city, is conducive to open and enthusiastic identification, especially on the part of individuals who come to feel that their happiness and that of their families is closely tied up with the neighborhood" (p. 365). This sense of neighborhood identity serves to shape a person's individual identity as well.

McMillan and Chavis (1986) attempt to define a psychological sense of community, as there are varied interpretations of the term. For them, sense of community involves four characteristics: membership, influence, integration and fulfillment of needs, and shared emotional connection. Each element, as imagined, is a confluence of certain factors to build a robust sense of community. Membership means people feel as if they belong in a community, and the converse of being an "other" also holds true. Physical boundaries such as gates delineate who belongs in a community and who does not, as do non-physical boundaries such as race and socio-economic status that might limit where people choose to live. Membership includes elements such as shared symbols (history, branding, culture, etc.), personal investment in the neighborhood, and a sense of belonging.

Second, influence means how someone can become meaningfully involved in the neighborhood, how their voice is heard, and how their voice influences others (McMillan and Chavis, 1986). Conformity can engender increased neighborhood participation, but it also can create a sense of sameness and groupthink. It could cause fissures if someone feels they lack influence (McMillan and Chavis, 1986). Third, integration and need fulfillment means how the neighborhood is satisfying needs someone prioritizes. That is the catch, as individuals experience and rank needs in differing manners, so there is no blanket solution. Ideally, strong neighborhoods allow someone to meet their own needs while also contributing to a common good to build up the neighborhood as well (McMillan and Chavis, 1986). Finally, shared emotional connection is how one creates and sustains a connection to a place. For branding purposes, this emotional connection is vital because it serves as a cognitive shortcut for decision making. If a neighborhood is labeled "bad," people do not have to investigate and think further – the emotional reaction is already there.

Cochrun (1994) notes that sense of community is largely a psychological, individual-level construct and is "the feeling an individual has about belonging to a group and involves the strength of the attachment people feel for their communities or neighborhoods" (p. 93). Cochrun (1994) details the relationship between community organizations, be they formal or informal, in creating a sense of neighborhood community. Germane to the public administration focus of this book, they argue that local public institutions such as governments and schools are physical spaces to foster a sense of community. In my findings, schools were primary ways adults made connections in their neighborhoods – though this is not a catchall.

One parent who lives in Baldwin Park in Orlando said he and his family chose the neighborhood because of proximity to good schools – and it is walkable, making it feel quite like the United Kingdom from where they moved. His children went to Brookshire Elementary, a public school in Orange County. "Even now we still talk about the Brookshire community because when we first moved here, we moved a long, long way and we were so happy to find a place where we settled so quickly, which we didn't expect because that's not what you think about Orlando." He said his family specifically chose a good public school because "I'd rather be in a good public rather than a poor private [school]." Many of their friends today are parents from Brookshire.

Alternatively, sometimes the school choice does not lend itself to creating a sense of community. A resident in the San Fernando Valley of Los Angeles said she and her husband picked their neighborhood because of the school. "There's this little elementary school right smack dab in the middle of my neighborhood, and we did some research on it. We looked it up on Great Schools (a school ranking website). It was fine. It wasn't amazing, but it was close and seemed to be on the rise as far as scores." In her area, though, there is school choice so it turned out her decision to choose a neighborhood based on the school was irrelevant in some ways.

In sum, a neighborhood sense of community relates to place attachment to reveal how invested someone is in building their neighborhood's social, economic, political, and cultural capitals. As McMillan and Chavis (1986) point out, it is a complex psychological process influenced by myriad factors. Sense of community can be packaged and sold to attract people to certain neighborhoods. Regarding brands, "what creates a sense of community around a brand is the sense of sharing meanings and belongings (associated identities). Sharing and belonging creates a sense of purpose for people" (Aitken and Campelo, 2011, p. 916). Collective ownership is a vital part of this sense of community, and we see this in neighborhood pride (see also Wherry, 2011).

As one branding expert told me, neighborhood branding is different from city branding, which also differs from corporate branding. Many corporate brands actively seek and develop a brand strategy, and increasingly places are doing the same (Zavattaro, 2013). With neighborhoods, he said, "I'm not sure if neighborhoods seek a brand. I think lots of brands simply emerge. I don't think a lot of them deliberately set out to be a brand in a commercial sense." More established communities in Central Florida such as Winter Park (its own city) and College Park (an Orlando neighborhood) have brands that "just evolved over many, many years," he explained. Even if a neighborhood does not strategically seek a brand, "you want something that is marketable to the outside world at the end of the day. The most effective marketing is word of mouth, through the people."

Influence of technology

Though word of mouth remains efficient, several people to whom I spoke mentioned the power of digital communication in creating a sense of community. I list it here as an important pattern to consider related to neighborhood branding and identity. Social media use within a neighborhood often attracts a younger demographic, and Johnson and Halegoua (2014) found those users often have children and are newer to the neighborhood so turn to social media to build connections with neighbors. The tools, though, are just that – tools. Residents expressed concern about being able to find the pages, fresh content, and how the pages are used, wanting less noise and more information (Johnson and Halegoua, 2014). A main reason for avoiding social media use within the neighborhood was unequal access, potentially leaving behind people not on the sites (ibid, 2014).

There are several ways in which technology can help or hinder creating a neighborhood sense of community. First, social media have changed urban planning and design processes. Social media are understood as tools to foster an information exchange and potential dialogue and include platforms such as Facebook, Twitter, YouTube, Pinterest, Snapchat, Instagram, and more. Social media "provides a medium to extend place-based interactions" by allowing people to quickly share pictures or videos (Evans-Cowley, 2010, p. 408). Social media also can create spaces for information sharing and participation by involving residents within the planning process by not putting a burden on them to drive to a public meeting (Evans-Cowley, 2010).

In an interesting study, Santala et al. (2017) use the phone app Untappd to map craft beer hotspots in Curitiba in Brazil. They studied where people checked in on the app and created heat maps, showing beer-drinking hot spots. From there, urban planners can use the data to create tourism-influenced "beer trails" showing people where and how to spend money and travel within the city along the newly created "trail." In turn, city and neighborhood officials can then market the beer trail and breweries to create buzz and investment within the place (Santala et al., 2017). (Where I live, there is the Central Florida Ale Trail, a physical map of nearly 20 breweries in the region. People can bring the map into the various bars for a stamp marking they have been to the place.) In sum, there is a relationship among social media, neighborhood branding, and urban planning.

One urban planner in my study expressed much the same sentiments. She is a planner with Tavistock, the company responsible for designing Lake Nona in Orlando. Laureate Park is a neighborhood within Lake Nona with colorful homes and apartment complexes. She said Laureate Park is a gigabit community with fiber optic lines to provide for the community's own internet and cable television. "The infrastructure to support technology and its users has changed planning," she said. For example, if a private utility company wants to come into the Laureate Park neighborhood, Tavistock placed conduits outside of the right-of-way easements, meaning that the companies do not need to tear up sidewalk and road-ways to install cable lines. In this way, Tavistock can sell access to other utility companies, saving both parties time and money.

In addition to the physical planning around technology, "with social media, one of the trends is master planned communities having these tweetable moments or these Instagram opportunities," she said. For example, a salon owner in Lake Nona wanted to create those photo-worthy moments so commissioned a local artist to paint angel wings on the side of her building. People now stop there to take and share photos, tagging the salon and the community on social media. In Laureate Park, there is a stained glass doll house that allows people to snap photos with a lake as the background. She said: "As a planner, we try to figure out what are those opportunities? How do you keep it classic so it [doesn't] just go away? That's been an interesting change over the last 10 years with how we plan spaces. You certainly don't want to have a place that's empty or rundown, and it's in the background of somebody's picture. It's free advertising."

Second, social media technology has been able to connect people in their neighborhoods as a means to share information and promote community events. Again, this is not a panacea and does have some drawbacks such as accessibility and ease of use (Johnson and Halegoua, 2014). I spoke to a representative from Nextdoor, the online platform designed for neighborhood interaction and communication. Residents can join Nextdoor and subscribe to feeds from what the algorithm decides is your chosen neighborhood. (For example, I can access the Baldwin Park/Audubon Park Nextdoor site but not, say, Thornton Park in Orlando.) She said Nextdoor has a presence in all 50 states and hosts 170,000 neighborhoods throughout the US on its website. "Our neighborhoods are everything from rural, urban to suburban. Neighbors are neighbors everywhere." The company's founder, she explained, is from Odessa, Texas, a relatively small town, but now lives in San Francisco with three children. "You've got Facebook for your friends, and you can stay apprised of what's happening around the world. They thought, 'What if we could give neighbors an easier way to connect?' and sure enough it was needed, and it's taken off."

She said the biggest use for Nextdoor is seeking recommendations from neighbors, as well as posting about crime, safety, and jobs. Additionally, Nextdoor has more than 3,000 government partners that use the site to communicate directly with residents in those neighborhoods. She said public agencies often use the site to push out information because of their broadcast model – people can post seeking help, recommendations, advice, etc., so they are broadcasting information. She said Nextdoor began "onboarding public agencies a year ago (2017) but the demand is huge. Agencies have their own pages, and they can push out messaging, and they can be really targeted. In [Hurricane] Harvey, for example, Houston is huge, and they were all using it during the hurricane. But what was really cool to them [public agency partners] is they could really cater information and really localize it, which makes it that much more engaging."

She gave me several examples of people coming together on Nextdoor, creating a sense of community in the virtual space. She spoke about Syrian refugee families in St. Louis, Missouri, who found assistance through posts on Nextdoor. Neighbors rallied 500 volunteers to assist with language lessons, lodging, transportation, and employment. A missing child was reported on the news, so neighbors turned to Nextdoor to help find him. A woman in Texas needed a liver transplant and found a match through Nextdoor. The Nextdoor executive said, "For me, Nextdoor has actually proved to be very valuable in elevating neighborhood identity in bringing people together and neighbors together around local issues. Stuff we see on Nextdoor would be hard to do without Nextdoor. I think that because it's your neighbors you're just really motivated to help out. I don't know of another social media platform where you'd be that motivated to raise your hand."

Springfield, Missouri, public information officials explained their neighborhood communications plan, which includes Nextdoor. They use a combination of social and print media in an attempt to avoid some of digital technology's exclusionary elements (Johnson and Halegoua, 2014). One team member said they have a presence on Nextdoor as an official partner agency so push information that way, as well as Facebook, Twitter, and Instagram. She said they also have a city-owned television station and a quarterly newspaper. "It's just a really effective medium, and people still like to read newspapers. It's nothing flashy like a perfect magazine. It's something that everybody knows how to use. It's free. They can pick it up, or they can get it in the mail. I think people really get a kick out of seeing themselves in print as well."

These sentiments regarding social media use align with research showing digital media can increase the strength of weak ties (Hampton and Wellman, 2003). Hampton and Wellman (2003) found that people using the neighborhood's networking capabilities and social tools felt more attached to the place, allowing them to meet others beyond their own block. More than social media, though, was time lived in neighborhood that really strengthened ties and allowed for more personal connection, so the authors are careful to say there is a balance because technology cannot solve all neighborhood issues (Hampton and Wellman, 2003).

Third, social media can create a sense of community within the neighborhood, but there are several challenges associated with the platforms such as access to technology and media literacy. Examples given from the Nextdoor executive illustrate this point. During crisis situations, social media are important tools for communication, but what happens if the power goes out or the phone battery dies? I am not suggesting that people in the interviews said social media can replace traditional forms of communication. Indeed, social media and other communications tools (press releases, websites, radio, television, magazine, newspaper, etc.) often work together. Social media in neighborhoods, in an ideal space, can create a sense of community by sharing information that brings people together. Instead, they did laud the benefits but also made sure to discuss challenges.

Said one member of the City of Orlando's CNR team: "We encourage the neighborhood organizations to have social media accounts because of the younger generation, the 24/7 connectivity, the need for information, or not to have to call someone to get info." What happens, though, is sometimes residents will use social media to post photos of car break-ins or other crime in the neighborhood. This image, then, can become prevailing.

The (as of this writing) president of the Baldwin Park, Orlando, residential association explained neighborhood social media in this way:

> *I think because it's new and it's easy and there are so many things that it can do, it has become a bit of a catchall and has lost some of its focus for people. You're getting, in my view, considerably less information that you really want to spend time with and more information about stuff that, it might be interesting, but I don't need to spend 30 seconds on the next tasty meal. I think in terms of what it's done to Baldwin Park, I think it's, despite Nextdoor and Facebook, it's significantly underdeveloped as a resource for the community in terms of the leadership of the community and my leadership. I'm sort of using that term for lack of a better way to describe that board that's responsible for the community association. It's underdeveloped because the people who are responsible in the management company see it as more work than they have time for. Instead of using it as a resource to widen the engagement and educate and promote, it becomes, well we would have to monitor this every day.*

He said the neighborhood does have a weekly email newsletter, which people have to sign up to receive. Those newsletters, while helpful, have people spend a lot of time scrolling to find information, he explained. Instead, he suggested each item in that newsletter could also be included on social media with a photo and link in case people do not wish to scroll. Perhaps some of that is changing, though it could be slowly coming.

The planner at Tavistock, speaking about the Laureate Park neighborhood in Lake Nona, said in addition to planning Instagram-worthy moments such as the glass house, social media can be another marketing tool to reach residents. There is a Laureate Park Facebook page, and other master planned communities might use cellphone apps that allow you to interact with others or pay for services. I asked her about Nextdoor, and she said, "I don't think Nextdoor has necessarily changed my profession. I don't know that it's changed the way I look at what I do or the neighborhoods I create. Like any other social media platform, it may go away. Technology and social media in general have changed the way you do your job."

One woman who lives in the San Fernando Valley in Los Angeles said she joined Nextdoor to connect with other neighbors because "we don't have a Facebook page or a neighborhood watch or anything." She said if there is a problem in the neighborhood or police activity, people tend to post it on Nextdoor to crowdsource what happened. She explained: "I think the Nextdoor app has been really helpful. It puts me in touch with different people in the neighborhood, my kind of extended neighborhood that I wouldn't have the opportunity to speak with. . . . Nextdoor is a blessing and a curse. At one point we had people talking about doing an armed militia in the street. For better or worse you learn about the crazies in your neighborhood."

Masden et al. (2014) studied Nextdoor use in three Atlanta-area neighborhoods, interviewing users after posting on Nextdoor to find study participants (I did the same in line with my study procedures). They found that people in their sample already tended to be active neighborhood participants through both formal and informal activities, such as clubs or block parties. In this sense, the platform was able to enhance or augment existing social ties while encouraging people such as renters to build new ones within the community (Masden et al., 2014). Further, users in their study reported enjoying Nextdoor's ability to have targeted, granular conversations rather than, say, a Facebook post that reaches the entire neighborhood rather than a particular subset. Self-curated, functional content was a top draw to the Nextdoor platform, yet challenges remain when the app draws boundaries that might not reflect the actual neighborhood geography or composition (Masden et al., 2014).

Feelings of safety and neighborhood identity

The second pattern to emerge related to neighborhood identity creation was feelings of safety. If someone feels safe where they live, they are likely to tell others. Safety serves as a cognitive shortcut for some when choosing where to live because it separates the so-called good neighborhoods from the bad. These labels are often difficult to change.

Safety for people is more of a feeling than something that relies on heavy use of crime reports. Typically, negative feelings of safety are associated with neighborhoods on the decline, showing potentially worn-down conditions (Austin, Furr, and Spine, 2002). This aligns with the "broken windows" doctrine that postulates that broken windows or dilapidated structures indicate little investment in the neighborhood so can spur similar destructive actions (Kelling and Wilson, 1982).

Put simply: safety is often a perception and something that is inherently part of a neighborhood's image, for better or worse.

Feelings of safety can increase someone's place attachment (Comstock, 2010) and often include formal and informal aspects (Sampson, Raudenbush, and Earls, 1997). Neighborhood watches or patrols are examples of formal mechanisms to demonstrate safety, while informal elements include self-monitoring or willingness to "help thy neighbor" in potentially dangerous situations (Sampson et al., 1997). For interviewees, they focused more on the personal side of safety rather than express communitarian aspects. In other words, people discussed more of their personal safety rather than an overall neighborhood safety.

I could have placed many of the findings in this chapter within the discussion of coded language and "Like Me" bias found in Chapter 6. People in the sample reported avoiding certain neighborhoods or gravitating toward neighborhoods with walkability and good schools. Most of the time, those latter neighborhoods were in wealthier areas of respective cities. These feelings are common and match with existing literature showing predominantly African American neighborhoods are perceived as more dangerous and uncivil compared with white counterparts (Taylor and Covington, 1993). This relationship is often tricky because neighborhoods thought of (brand image) as unsafe or blighted often do not receive the same level of municipal services compared with neighborhoods with higher socio-economic status (Taylor, Shumaker, and Gottfredson, 1985). This makes healthy neighborhoods a key focus on many local governments wanting to curtail images and realities of unsafe neighborhoods.

In terms of place branding, there is a link between public spaces and overall individual and community well-being (Beck, 2009). As Beck points out, access to these kinds of spaces is not only a planning issue but also largely an equity and social justice issue. Poorer communities might lack open green spaces and safe parks (Beck, 2009). Cities with spaces the public perceives as safe can promote those as value-addeds to attract tourists and even local residents to the areas (Chan and Marafa, 2016).

Feelings of safety defined

Perhaps not surprisingly, safety and fear of crime are contested in the academic literature. Shirlow and Pain (2003) explain that fear of crime "covers the wide range of emotional and practical responses to crime and disorder which individuals and communities make" (p. 17). Fear of crime historically rears its head when related to some kind of panic or widely publicized crime (Shirlow and Pain, 2003). Rhetoric, images, and discussions of fear often are coopted by powerful groups in an effort to make someone the "other." They use the example of terrorism and othering of Muslims around the world. Similar policies such as stop and frisk that allow for detaining people seemingly deemed suspicious also serve similar effects – to other a minority group while showing those in power that something is being done to allay concerns regarding fear of crime (Shirlow and Pain, 2003).

Hutta (2009) critiques both the feelings of safety and fear of crime constructs, arguing similarly to Shirlow and Pain (2003) that they are means of oppression and an exacerbation of power relations between the marginalized and dominant groups. He lists policies such as zero-tolerance crime manifestos that usually target minority populations, such as the US saw after the September 11, 2001, terrorist attacks or general stop-and-frisk policies. Fear of crime and feelings of safety become discursive practices meant to manipulate people into targeting those who do not look like them and "have become generic terms that allow for making statements about people's subjective feelings as well as about what seem to be the central problems of the community" (Hutta, 2009, p. 253). Boundaries further separate the so-called safe from unsafe, manifesting in practices such as redlining and gated communities where physical barriers are erected to showcase feelings of safety.

One branding expert explained that economically focused neighborhood branding and marketing campaigns that promote a positive quality of life and good schools, for example, also implicitly sell safety. Using an economic focus, "you're only going to attract doctors, professors, and engineers if it's a nice place to live with good schools and it's safe," he said. He lives in Orlando's Baldwin Park neighborhood, choosing it for the good schools, walkability that reminded him of home in the United Kingdom, and safety. "I don't like gated communities because to me they're not real. Some people say it's less safe (without a gate). I don't believe that. Anyone can jump over a fence."

When people brought up safety during the interviews, I asked them to explain what they meant. The following are some of the responses:

A front porch light is a sense of safety. It's not so dark. It's feeling safe in my neighborhood, and I kind of look at the whole thing. Walkability is great, but I have grown kids now but I want to be able to know the good schools.
– City of Orlando neighborhood relations team member

It is safe just about any time you drive around the lake. Even at night you will see people walking around the lake.
– Baldwin Park, Orlando, resident

Safe means if I choose to leave the house unlocked it's probably okay. It's not a good idea, but it's probably okay. I'm sitting in a garage apartment now, so when I come up here I do make sure the front door is locked because if someone rings the doorbell I won't hear it. If someone walks in I won't know it. . . . It's just a little caution. Safe is my possessions are safe, my property is safe. Safe is there's not going to be theft, there's not going to be crime. Does that happen? Yes. Did we have car break-ins one week ago? Yes. Three blocks away. Do I leave sunglasses in my car? No, I don't. I remove temptation.
– Celebration, Florida, resident

I think that is a misconception. I don't know that you can really say any neighborhood is safe because Audubon Park has crime. It's not murder. We have

car break-ins at night. We have bikes being stolen. I think in terms of a safe neighborhood, I think it's definitely one that has a sense of community and has neighbors that look out for each other.
– Audubon Park, Orlando, resident

I mean, my car was opened and rummaged through at my place in Dallas, but I've never felt unsafe, and maybe it's just because I grew up in areas that I knew were unsafe. As an adult I've never been nervous walking around my neighborhood at night, but there are people who have been. If I'm standing at a dark station [waiting] to get on the train, my coworkers won't let me take the train after dark. Homeless doesn't mean dangerous.
– Dallas, Texas, resident

Urban regeneration that can be packaged and marketed as the latest hip and cool neighborhood or area is often code for furthering othering of already marginalized populations. Bennetts et al. (2017) examine the link between the built environment and feelings of safety, specifically finding that there are elements of new urbanism-style design that promote a surveillance culture – or people watching, however you like to see it. Jacobs (1961) articulated as much when she described living streets that had people watching, looking to see who belonged and who did not. In their study, Bennetts et al. (2017) found that mixed-use–style buildings that encouraged sitting outside and mingling made people feel safer compared with buildings that shut down in the evenings. Feelings of safety in the urban environment are a complex interaction of movement, activity, familiarity, and maintenance (Bennetts et al., 2017).

As quotes from the interviewees demonstrate, feelings of safety often are related to perceived neighborliness and cohesion. People spoke about helping one another, looking out for one another akin to Jacobs' (1961) findings. Strong neighborhood social networks not only support place attachment but also feelings of safety and neighborhood engagement (Unger and Wandersman, 1985). People use cognitive mapping to identify neighborhoods in which they feel safe so will intuitively avoid what they have labeled as "bad" neighborhoods where they might feel unsafe (Unger and Wandersman, 1985). In this way, people can recognize those who "do not belong" to potentially mitigate threats. Jacobs (1961) articulated similar principles as well with neighbors looking out windows for anomalies.

A Baldwin Park, Orlando, resident explained that feelings of safety sometimes relate to artifacts within the community. The resident said police officers often patrol the neighborhood, yet "people will say, 'I've never seen the police in Baldwin Park ever.' Well, I don't know where you live, but they're here." During the summer, the association will "hire off-duty police to patrol the parks and the pools and things because it's where there is most likely the frat boys drinking and cussing and making a lot of racket and detracting from it being a family place to gather." He said the community also added security cameras in public places such as the pools and parks to increase feelings of safety. This is not uncommon in a surveillance society (Foucault, 1977).

Gendered elements

Despite a surveillance culture and elements in the built environment meant to increase feelings of safety, there still remain concerns when it comes to gender. Largely, women in the study reported thinking more about safety compared with men. This is not to say that men did not mention safety; they did, of course. But women were more acutely aware of neighborhood image related to feelings of safety. Women often are told to avoid certain places after dark, travel with friends, use their keys as a weapon, and dress appropriately. Feminist geographers examine the relationship between the built environment and feelings of safety (Sandberg and Ronnblom, 2015), arguing you cannot uncouple these constructs when trying to understand the built environment (Koskela and Pain, 2000).

Women often report higher levels of fear of crime and are statistically more likely than men to be victims of sexual assault and violence (Tandogan and Ilhan, 2016). Women report the highest fear of crime in the following spaces, especially when alone at night: quiet roads, dark subways, poorly lit streets, empty parks, abandoned buildings, subways, and urban spaces without mixed-use value, to name a few (Tandogan and Ilhan, 2016). Most women reported not walking alone at night in these public spaces to reduce their risk of becoming a crime victim (Tandogan and Ilhan, 2016). Pain (1997) similarly finds that women's fear of crime often manifests in avoiding public spaces, thus curtailing areas in the world where women allow themselves to be present. Fear of crime and safety in private space also trickles into public space and vice versa, so this is clearly a complex issue (Pain, 2001).

The following is a sampling, then, of gendered responses to feelings of safety:

> *I come from a city in which street sexual harassment is a problem. I have that in mind. I want to walk and not be sexually harassed.*
> — *Tucson, Arizona, referencing Mexico City, resident*

> *Ideally, like in an ideal world, I'm a woman so I think I have different opinions than my boyfriend for what safety means to me. Walking home alone at any time of day, day or night, winter or summer, and feeling like I'm going to leave this one- to three-mile radius, and I'm going to get home safely. Sometimes I do feel that that's not necessarily the case.*
> — *Baltimore, Maryland, resident*

> *It's hard to divorce my personal from my professional take on this. I'd say it's really, there's two aspects of perception of safety and the reality of safety are very real things. [Washington] DC has a decent amount of crime, and I'm cognizant of that, especially for my wife. It doesn't necessarily have the same stigma as the data would bear out. I think because I love cities, and I think about safety, I feel comfortable being in most neighborhoods in most times, places, and frankly being an able-bodied man in his mid-30s certainly helps, too, in terms of the safety element. I think a lot of people, though, are less*

rational, which is totally fine in terms of what makes them feel or not feel safe. And there are certain areas where people don't feel comfortable. You hear people don't feel comfortable going to downtowns in certain places.

— Community development expert, Washington DC

It's very safe, too. As a single woman that was important to me. It's not like a gated community. It's open but, kind of, the people in the neighborhood are really calm. It's not like a party area. Probably professionals around my age, some married couples with young kids.

— San Diego, California, resident

She loves it here, but the other part of moving where we live now is that the physical living space that we have now is a really nice living space. It's attractive. It's quiet. It's private. She was, when she bought this place, a single woman, and the idea of feeling private and more secure coming home alone at night was very important.

— Montclair, New Jersey, resident

I'm a runner so not feeling like I'm going to get grabbed or assaulted when I'm running. It happens a lot in Center City, Philadelphia. A lot. Safety is for me, it's the little things. Like in the wintertime being able to go out to your car and warm it up in the morning and don't worry about it getting stolen. [My] sister is 27 and wanted to go for a walk and there was a nearby mugging. I like my heels for work but I like to be able to move quickly. I keep all my shoes, like, in my car and this way I have all my options available depending upon what I'm wearing.

— West Mount Airy, Philadelphia, Pennsylvania resident

The foregoing quotes demonstrate how public places and geographies can be spaces of oppression for women and other marginalized communities (Valentine, 2007). This oppressive relationship becomes even trickier when intersectional identities are taken into account (Valentine, 2007). Women generally report increased fear of crime when walking alone at night or while home alone, and stranger harassment only increases these fears (Macmillan et al., 2000). Neighborhoods can market and brand themselves as safe, but people themselves are the ones who experience elements related to safety. Safety becomes a social construction relative to the interplay of myriad personal and public factors.

Concluding remarks

Both sense of community and feelings of safety are something cities and neighborhoods alike can brand and market. These could be positive or negative associations, with negative brand images difficult to overcome (Avraham and Ketter, 2008). Regarding safety, Avraham and Ketter (2008) note that places can use a "come see for yourself" strategy (p. 198) to show people that perception does not match reality. In Madisonville in Cincinnati, Ohio, they did this when the perception of the neighborhood was one of malaise and unsafe conditions. An executive with the redevelopment corporation told me that Madisonville was statistically the sixth safest neighborhood of 52 in Cincinnati in 2017, but he said the perception of the area is one of high crime and crimes of opportunity (like stealing items from cars or open garage doors). "For a while the [branding] effort was, how do we convince people this place isn't as unsafe as they think it is?"

Both feed into overall reputation management for the place. Places with reputations seen as unsafe, for example, might suffer economic and social loses (Coaffee and Rogers, 2008), and uncertain political landscapes also can have the same dulling effect (Zavattaro and Fay, 2019). Place branding can help create a sense of community through shared expressions and image use (such as a common logo or slogan, along with shared values), and these shared elements can create legitimacy – or abandon it – when used together (Sevin, 2011). Neighborhoods that can promote a sense of community through formal or informal means (events, block parties, associations, etc.) can heighten place attachment and even a sense of self (de Azvedo et al., 2013).

3   Government-neighborhood relations

*All those neighborhoods around downtown, we formed what's called a coterie,
[a] citizen-driven process. These are all names of neighborhoods that have been
around for many years, so they formed coteries to work with local government to
reimagine neighborhoods and reinvent them.*

*— Convention and Visitors Bureau executive,
Lafayette, Louisiana*

A coterie is a group of people with a shared common interest. The executive
explained how residents in Lafayette's older neighborhoods came together to
work on revitalization issues. He said: "One of those neighborhoods had not had a
new house or anything built in 50 years, and about four years ago there was a loft-
type apartment complex that was actually for lower- to moderate-income folks
that started pulling a lot of fear from the neighborhood. They fought it like crazy.
Now it's been a key piece to the revitalization of that area."

This chapter is about the nexus among residents, governments, businesses, nonprofits, and other stakeholder groups creating vibrant neighborhoods. My training is in public administration, and I happen to study place branding at the local
government level. This chapter brings those interests expressly together to highlight how working relationships affect an overall neighborhood brand identity and
image – and vice versa. Sometimes, as it was explained to me, some neighbors
perceive they are lower on the priority list for services from their city government,
leaving some anxiety and stress when it comes to healthy neighborhoods. The
power differentials are often real, rooted in political and socio-economic power
imbalances (Goetze and Colton, 1980).

In this chapter, I highlight two major aspects that stood out during data analysis:
neighborhood governance (usually an association), and the government and community development elements of neighborhood branding.

Neighborhood governance

By neighborhood governance, I am detailing mostly involvement with homeowners' associations (whatever the nomenclature may be). People to whom I spoke
described getting involved because they had a passion for their neighborhoods,

for good governance, or sometimes both. No matter the motive for joining, they wanted to make a difference in the neighborhood through volunteering. I use Chaskin and Garg's (1997, p. 632) definition of governance to understand these relationships: "the creation or adoption of mechanisms and processes to guide planning, decision making, and implementation as well as to identify and organize accountability and responsibility for action undertaken." As can be seen reading the definition, governance is complex because it is both a process and structure (Chaskin and Garg, 1997).

Yates (1972) traces the history of neighborhood governance in the US, noting that American ideals and machine politics often thwarted neighborhood self-government. What spurred neighborhood self-government, according to Yates (1972), was the idea of decentralizing certain government and engagement functions directly to neighborhoods, such as health clinics, school boards, and poverty negation programs. There are four major challenges for establishing meaningful neighborhood governance: costs of organization and participation, community conflict and tension, disagreement with city hall, and general politics. "This means that serious participation is likely to occur only when neighborhood government programs offer visible rewards and work to solve concrete problems" (Yates, 1972, p. 214).

Continuing with the historical development of neighborhood governance, Chaskin and Abunimah (1999) explain that various US federal programs required municipal involvement of some kind, especially related to neighborhoods. One they mention was Model Cities Program, which was established in 1966 under the Department of Housing and Urban Development (Strange, 1972). The Model Cities Program was meant to leverage public and private resources toward combatting neighborhood blight, so program responsibility went to the mayor and city council rather than directly to the residents, though residents should ideally have direct access to decision makers and processes to ensure meaningful involvement (U.S. Department of Housing and Urban Development, n.d.). Model Cities received special funding from the government for five years to address urban blight, education, building dilapidation, economic development, job training, and more.

This division in service provision between the local government entities and residents caused both tension and confusion (Strange, 1972). Poor and minority communities often were left out of participatory mechanisms, which were slim to begin with related to these programs (Strange, 1972). Local governments even went so far as to change or eliminate participatory requirements to regain control. These kinds of actions are not uncommon and often lead to distrust between residents and local governments (Chaskin and Abunimah, 1999).

The 1970s in the US saw a pushback against efforts to thwart participation in neighborhoods, and even municipalities encouraged decentralization of certain governance functions to newly created neighborhood associations and organizations (Chaskin and Abunimah, 1999). Governments themselves created participatory mechanisms such as citizen advisory boards and annex city halls within neighborhoods rather than only one main building (Chaskin and Abunimah, 1999) to further build connections.

I spoke with city officials from Springfield, Missouri, in the public information division who implement many of the neighborhood engagement practices Chaskin and Abunimah (1999) detail. They explained their neighborhood focus revived in about 2012 when the city manager created a position of director of public information and civic engagement. The team created the Neighborhood Advisory Council, recruiting representatives from the city's 16 registered neighborhood associations at that time. "At first it was a little rocky because nobody understood what the role was or what could be done with that," said one team member. She said it took at least two years for the initiative to really get going thanks to "an amazing engagement group" that now meets quarterly to discuss neighborhood issues with city officials and suggest priorities to staff and council members. "I would say their opinions have a lot of sway. Not all of them get funded, of course, but they feel very empowered. And to me that's the ultimate in civic engagement."

In another example, I spoke to a resident of the Millers Bay neighborhood in Oshkosh, Wisconsin, who was one of the founding members of the neighborhood association. She got involved with creating the association because of her passion for beautification projects. She has served as president for four years and vice president for two. She said the association has a neighborhood plan based on the city's own strategic plan, focusing on areas such as utilities, economics, transportation, and housing. She has worked with the local government and other entities to earn grants to improve landscaping at the local elementary school, a project she called "unprecedented" because of collaboration among her neighborhood, another neighborhood, the school board, and city. She said, "Once you have a recognized neighborhood and you have a neighborhood plan you have a little bit more, I don't want to say clout, but the city partners with you then on your vision."

What these examples show is the interplay among neighborhood residents, city governments, nonprofit organizations, and businesses when it comes to issues of neighborhood governance. Chaskin and Garg (1997) explain that many neighborhood initiatives aimed at neighborhood change (rebuilding a park, community centers, etc.) are funded through grant-making agencies, introducing a new set of stakeholders into the mix. Neighborhood governance is challenging because neighborhoods are not actually independent agencies; they are part of a municipal whole (Chaskin and Garg, 1997). The next section goes into more detail about this complex relationship.

Government and community development elements

As noted earlier, the biggest challenge when it comes to neighborhood governance is finding people to spend the time and energy to serve on associations. City of Orlando CNR team members told me this is the foremost hurdle in their plan to develop sustainable neighborhood associations: finding people to participate. The CNR department assistant director said that "some [associations] are super active, others may meet twice a year, but the reality is they have a community network. They have leadership in place." Another member of the CNR team said there are about 350 neighborhood associations in Orlando, and "we have a relationship with about a third of them."

All the CNR team members said the purpose of their office is to be the face of City Hall for residents. They work on behalf of the mayor, who cannot attend all neighborhood happenings and events. Their job is to make connections with people, serving as their direct link to local government. Said one team member: "What we do is we work with our neighborhood associations throughout the city, and residents, but primarily associations because we are a small team. We work through the association boards through to the residents." The office is called Communications and Neighborhood Relations because they push out directly to the neighborhoods information the communications side produces. Another team member explained, "Most cities don't have a dedicated neighborhood team. When budget cuts and things happen, police isn't going to go away, public works isn't going to go away. This isn't a need to have, it's a nice to have."

The CNR team agreed that their role is vital to simplify government speak, giving several examples of how this happens. In the first example, they said a working-class neighborhood was having issues with a roadway without adequate lighting. The lighting problems were just outside the neighborhood boundaries so residents were not sure what to do or who to call, so the CNR team facilitated working with the local utility company to install 19 lights to make the area safer. In another example, they explained working with the internal planning department to simplify communications about neighborhood projects. One member said the planning department was receiving criticism for not communicating well about projects that involved construction and noise in the neighborhood, "so they were being kind of reactive as opposed to proactive to get the information out." Instead of communicating in legal notices or complex public hearings, the CNR department transformed the mailers into a color postcard with the essential details. "We said, 'Let's make this something that people can actually see and be able to participate in.'"

With another example, CNR team members said code enforcement often is confusing for residents with the complicated rules and regulations involved. In response, the CNR team created an easy-to-read booklet to give to residents at public meetings explaining the code process, what to do when a violation occurs, how to report a violation, and even examples of what constitutes a code violation. Code officers can even give out the publication to people to simplify the process, explaining it in a concise manner. Said one member, "One of the biggest things I tell them [internal departments] is you have to make sure what you're trying to communicate is in a language people can understand."

Ideally, neighborhood association leaders are better connected to City Hall through this department because they have assigned to them a CNR team member. Commented one member, "I think that's what we also bring to the table in helping neighborhoods feel like they are part of the family in the city." I spoke to the leader of the Baldwin Park neighborhood association who also explained this connection with the city, detailing some of the complexities he and the association face with the various levels of government involved within the neighborhood.

He said there are various governing entities within the neighborhood, including the residential association, commercial development district, the property management company, local businesses, and two local governments, Orlando and Winter Park, a city neighboring Orlando that houses the dog park along Lake Baldwin. He said, "Needless to say it creates some obstacles that often make it difficult to explain that you don't need to complain to the residential owners' association about the commercial people blowing dirt off their sidewalks at 7:30 (in the morning) because the residential owners have nothing to do with it."

Baldwin Park is a master-planned community with strict building codes, landscape codes, and aesthetic limits. He said the neighborhood plan regulates paint colors and housing style. People can, of course, appeal decisions to the residential board but the city guidelines are quite strict for a reason. He gave this example: "Someone in 2004 moved here from Arizona, buys a new home and is truly bothered by the amount of sunlight that comes in and fades their carpet. So they get window tinting for the windows only to learn that window tinting is not allowed, and they object and say, 'I had this in Arizona, and it reduced my energy bill; it saved my carpet so why can't I have it here?' Well, you can't have it here because the guidelines say you can't have it here."

Because of the governing complexities, he described sitting on the board as "hurry up and wait" because the structure is "both an asset and a major hindrance." He gave an example of houses with portable basketball goals in the driveway or along the sidewalk. He said the neighborhood guidelines indicate if someone wants a portable goal, it needs to be cleared with the Architectural Review Committee within the neighborhood. And if it is portable, the goal must be put away after each use and not left in the driveway or along the sidewalk. He said, "The upshot of it was the board spent probably six months and multiple hours of board member time literally establishing how many homes had these basketball goal issues. What we also knew was that no one was complaining about them so the board was sort of wrestling with, 'we're responsible for administering this set of guidelines so how do we deal with this?'"

To generate an answer, he and the board reached out to Orlando Police asking what officers might tell kids playing basketball using a portable hoop in a neighborhood alleyway. "Well, we tell them to watch out for cars" was the answer from the police, indicating this is not a major concern. The resolution was to enforce existing rules of putting away the portable goals to align with neighborhood standards. Yet the various layers made the process time-consuming, as nothing was decided arbitrarily. "It's sort of a classic why do we have this rule if it's not enforceable?"

City of Orlando CNR team members also discussed how neighborhood relations relates to neighborhood image. It is sometimes perceived that wealthier neighborhoods receive more or better services compared with lower-income counterparts. Said one team member, "Some neighborhoods, they call and it takes a little bit so I feel like we try to help fight for that little guy and have those neighborhoods get the same love that other neighborhoods get in the city." There could be a perception that wealthier neighborhoods will get better services, so CNR team members try to alleviate this neighborhood image perception by responding to everyone. For instance, many of the "hipper" neighborhoods in Orlando have installed public art on street-side utility boxes. The CNR team is trying to bring this same idea out to neighborhoods on the west side of town to help them create their own identities through local public art.

Feelings of disconnection among lower-income neighborhoods are found within the scholarly literature as well. Lovrich and Taylor (1976) examined white, black, and Mexican-American neighborhoods in Denver, finding that black and Mexican-American residents were more likely to report being dissatisfied with government service provision. White respondents in their study also agreed that they were likely getting better services than minority counterparts (Lovrich and Taylor, 1976). Not surprisingly, these kinds of findings are more granular than a simple yes-no response, as van Ryzin (2004) found when studying expectations of local government service delivery in New York City. He used expectancy disconfirmation theory to test satisfaction with government services. The theory states that individuals form pre-judgments about service-level expectations before entering into service provision. This is where branding becomes important, given reputation of the organization is tied to service expectations.

Disconfirmation can be positive or negative if expectations are exceeded (positive) or not met (negative) (van Ryzin, 2004). He found that disconfirmation in either direction played a significant role in how citizens shaped expectation of service delivery. Germane to this project on neighborhood branding, he argues that (2004, p. 445, emphasis added):

> *Expectations may reflect memories of past performance in the jurisdiction or an experience living in or visiting another urban area. Media coverage of local government issues may be another source of influence. Yet another may be public information campaigns or other communications, such as official slogans or logos, that affect how citizens think about municipal government and the quality of urban services.*

Citizen satisfaction measures, though, can be problematic given they might not reflect actual service outputs and outcomes because of incomplete counting (Brudney and England, 1982). Related, comparing across neighborhoods might yield faulty results when other elements such as socio-economic status are not taken into account, showing the danger of blanket survey result reporting (Brudney and England, 1982).

Concluding remarks

How does this relate to neighborhood branding and identity? I asked people why a neighborhood would want an identity and what it could do for the place. Here I focus on what government leaders, residents, and community developers had to say when it comes to creating an identity for the neighborhood through an open, collaborative process. Chapter 1 detailed much of this, with the Neighbor-Works America example, for instance. Here the focus is on how the entities can work together to shape identity. I should note there was a divide in how important this was among interviewees. Professionals (government employees, community developers, Realtors, etc.) thought a neighborhood brand was critical while residents did not seem to pay a brand or identity much mind, unless it was used as a negative heuristic (as in, do not live in Neighborhood X because it is a so-called bad neighborhood). This finding calls into potential question the investment in neighborhood branding initiatives that often are difficult to see immediate results and return on investment.

The following are some snippets from residents about neighborhood identity:

> *There's a reason people want to maintain their identity. It helps with community, it's supporting local businesses. It's a sense of home.*
> — *East Riverside, Austin, Texas, resident*

> *I think that all neighborhoods and societies, they have their roots, that's where everything comes from. Local government, specifically in the US, should catch those values that might resonate with more people, so I don't think it's possible for a local official to shape an identity that doesn't exist from anything. They can echo voices or aggregate voices of the area. . . . An official cannot create the identity, but I think that the policies or decisions or ads can resonate with the grassroots of that area.*
> — *Tucson, Arizona, resident*

> *Why a neighborhood would want it (identity), I think it becomes a rallying point, a point to bring people together. It's a community builder. So that goes into branding and labeling.*
> — *Celebration, Florida, resident*

> *I don't think everybody is necessarily looking for the same thing in where they live. If you don't want to incorporate where you live into part of your identity, you move. They're taking advantage of a resource that's not really their community because they don't see it as part of their inherent identity. This is almost like a chicken-egg deal: do [you] incorporate it into your identity because you already see a shared value system, or does living there instill a sense of a shared value system because you see it as a shared sense of identity?*
> — *Normal Heights, San Diego, California, resident*

What is interesting about these sample quotes is they show a tension in how neighborhood identity (and associated images) are created. The Tucson resident, for instance, argues that a government entity cannot create a brand from nothing. This aligns with arguments that place branding fails because consultants come in and tell a place who and what they are instead of engaging in meaningful stakeholder engagement practices (Kavaratzis and Hatch, 2013). Others talk about how self-identity is intertwined within the neighborhood. This means that branding and marketing efforts will naturally appeal to those interested in those neighborhood characteristics if they match their own. If someone wants something walkable, that kind of marketing will appeal. If not, that marketing is a non-starter.

Eshuis and Edelenbos (2009) find that neighborhood branding should involve meaningful co-production because of its complex nature. A collaborative approach is best to ensure buy-in before branding processes become an open power grab. This is important because the brand identity (what the neighborhood creates) must match reality (what is really available in the neighborhood). In their study, Eshuis and Edelenbos (2009) argue that branding should match what is going on, good or bad, and give the example of the Katendrecht neighborhood in Rotterdam, the Netherlands. The neighborhood's slogan of "Can You Handle the Cape" matched the geographic realties but also made note of crime incidents in the neighborhood (Eshuis and Edelenbos, 2009). Changing neighborhood reputations, though, takes time and might not even materialize because perception is reality (Dekker and Varady, 2011).

Now to sample some quotes from professionals, such as government officials, Realtors, community organizers, and economic development leaders:

What I see here in a smaller town like Lafayette is that clinging to the identity of the neighborhoods, which if successful, I think, makes our job easier, where I can go and say if you're coming to Lafayette, here's what you can do in Freetown, River Ranch. Los Angeles is doing a very good job of that right now, sort of branding by neighborhood. . . . From a branding standpoint I think it's very complimentary.

– Convention and Visitors Bureau executive,
Lafayette, Louisiana

When you strengthen neighborhoods, you are strengthening the people who are there. Most people are conservative about what they want to have happen where they live, so if a city needs to respond to growth pressures there's tensions between identity and growth but at the same time experience can change things.

– City of Pittsburgh urban planner

How do we make ourselves interesting to people? And by selling ourselves, what does that even mean? It means we want people to live here.

– Madisonville (Ohio) Community Urban Redevelopment
Corporation executive

We are most successful in places where these efforts result in really robust collaborations. The bigger result is that organizations came together around a shared purpose and working collaboratively, that's what going to have a bigger long-term effect.

– NeighborWorks America executive

> *I would actually say the real Orlando is, and this is what I said to people back in Britain, I compare it to Birmingham. It has strong neighborhoods. It really is a strong collection of small villages with strong identities.*
> *– Orlando, Florida, branding expert*

These sentiments are similar to the residential take on branding but differ slightly in scope and impact. The NeighborWorks executive echoed findings from Eshuis and Edelenbos (2009) regarding the importance of collaboration. Oftentimes, though, branding agents fail to close the loop, fail to tell people in the collaborative how their input was used to create the brand strategy. NeighborWorks, however, does this, said the executive. Ideally the collective can retool and retune strategies as the neighborhood changes, constantly responding to feedback and data.

What some of the responses have in common is either an explicit or implicit economic measure of success. The seemingly ideal neighborhood brand is one that can attract investment, either in new retail, residents, or tourists. An economics-only measurement perspective becomes shallow when it comes to place branding because of the stakeholder complexity involved, as each group has its own ideas of success (Zenker and Martin, 2011). An alternative, for instance, could include content analysis of social media comments and influence networks (Andehn, Kazeminia, Lucarelli, and Sevin, 2014), where visualizing networks allows one to see a brand's reach and potential (Sevin, 2014).

Taken together, resident and expert perceptions regarding the importance of neighborhood branding shed light on meaning-making within the decision-making spaces related to branding projects. Sometimes place branding and marketing that tries to make a city or neighborhood stand out actually generates homogeneity (Renn, 2019) because there is a desire to "keep up with the Joneses" and imitate like-minded organizations (Fay and Zavattaro, 2016). Therefore, neighborhoods and cities may want to consider focusing on their place branding goals and purposes, as the NeighborWorks America representative suggested to me. When they consult with neighborhoods, he said they ask the neighborhood team to focus on priorities and values, realizing each stakeholder group will have its own unique goals and outcomes. "We often find that the strategies we develop, some of them are just core neighborhood revitalization strategies."

Getting people in the same room can yield big results, though the process takes time (Eshuis and Klijn, 2012), but the results often produce fruitful branding strategies.

4 Nostalgia and idealism in neighborhood branding

I know like four neighbors around me, and beyond that I don't know a lot of the people who live on my street. I recognize them. I wave to them. I'm cordial with them. I don't know their names or anything. I am friendly with my neighbors, but I am not friends with any of them. I'm a firm believer in the whole 'it takes a village' thing, especially when you have kids.

— San Fernando Valley, California, resident

While writing this, the United States and the world are going through a crisis of identities – progressive idealism versus regressive protectionism. The 2016 election of Donald Trump as president illustrates this problem nicely. Candidate Trump campaigned to "Make America Great Again" (MAGA), and though there is no clear definition of this trope it seems to mean a return to traditional, pastoral values that themselves were never fully realized in a society built in inequities (Wilkinson, 2018). Describing the MAGA slogan, Wilkinson writes (2018, p. 5), "It's punchy, but also vague and capacious, with enough room for anyone to imbue it with meaning."

For many, MAGA is about nostalgic reflections on an idealized version of an America that has not been the reality for many. Nostalgia is a powerful force in marketing and branding practices because it is all about an emotional connection (Hunt and Johns, 2013), which is critical to the success of product or place brands. Nostalgia, with roots in Greek medicine and thought and often referring to home-sickness, is defined by a sense of loss and estrangement (Turner, 1987). Turner (1987) explains four characteristics of nostalgia: historical loss and decline associated with a "departure from some golden age of 'homefulness,'" loss of moral certainty resulting from a values collapse related to religious fracturing, reduction in feelings of self and autonomy, and decline in personal authenticity. In the US political climate as of this writing, those feelings are in play, and those feelings manifested in the interviews for this research. Though interviewees might not have used the word "nostalgia" directly, many of them wished for simpler times and neighborliness that might not have existed prior.

Nostalgia defined

Nostalgia as a concept has moved from its roots in medicine to a social construct. While there is more social mobility, there might be less connection to a physical house or home, but there is a strong sense that nostalgia for thoughts, feelings, emotions, and actions can help moor our personal identities in times of potential turbulence (Havlena and Holak, 1991). "A nostalgia narrative is an imagined story of the past that deliberately selects certain elements from personal history while excluding others to construct a version that is more favorable than the reality. People weave a nostalgia narrative when they sense that their attachments to a place and their future in a place are under threat" (Ocejo, 2011, p. 287). Marketers realize the power of nostalgia narratives so use related theming to sell products and places, especially to older generations that might idealize a so-called simpler time (Havlena and Holak, 1991). The trouble is there is often authentic nostalgia, but there also is pseudo-nostalgia, drummed up usually by media and tourist attractions (Hunt and Johns, 2013).

Kasinitz and Hillyard (1995) explain that every generation has some kind of nostalgic thoughts or feelings for the past. For instance, the progress that was felt after World War II declined, replaced by feelings of unrest beginning in earnest in the 1960s. Nostalgia narratives in the past focused on historic preservation efforts, but more contemporarily nostalgia is found among working-class residents who feel left behind (Kasinitz and Hillyard, 1995). Sometimes it is easy to dismiss nostalgia as a simple longing for the past, but Pickering and Keightley (2006) argue this denies the power of meaning and time. Yes, there are problems with historical imagination, nostalgia as a capitalism booster, and misinterpretation of a past reality through rose-colored glasses, so to speak. But on the other hand, nostalgia can help revisit the past and learn from it to change our modern experiences. Invoking the past in a mythologized manner often leaves nostalgia as an easy scapegoat for historical inconsistencies, but often nostalgia can serve a purpose in helping us understand present, modernist conditions (Pickering and Keightley, 2006). Put simply: nostalgia can serve an individualized purpose in a collectivized setting. Nostalgia means something to someone in context.

When it comes to place branding and marketing, nostalgia is a powerful concept. In their study of stakeholder perceptions of the Santorini brand, Lichrou, O'Malley and Patterson (2010) found many subjects wanted to preserve an "island character" that hinges on preserving culture, architecture, and customs in the face of rapid development and change. Similarly, Eimermann (2015) studied marketing strategies related to the Swedish countryside, undertaken to play on nostalgia for greener, simpler times away from growing European cities.

Watson and Wells (2005) examine the rise of nostalgia on Poppy Street in London, which changed thanks to regeneration efforts that pushed out many of the locals. Even though forces of gentrification are taking hold of the neighborhood under study, business owners on the street largely were left out because "the area is perhaps too shabby to attract them" (p. 19). During their interviews, Watson and Wells (2005) detailed nostalgic sentiments from locals concerned about the influx of immigrants and refugees to the area, many of whom whites portrayed as "others" helping exacerbate problems of decline and big-government support. Nostalgia in general tends to erase negative memories of the past, and the scholars found such in their study – as I did herein. For example, shopkeepers spoke about crime and safety in the past, yearning for a time that was in their estimation safer that contemporary times despite crime statistics showing decreasing crime trends. This was evident in my interviews, too, especially when people spoke about neighborhoods where they grew up and could go play in the street unburdened.

Urban planning has not escaped the harkening for nostalgia. Today, neighborhoods want mixed-use spaces that are walkable and encourage public gatherings, embodied by the new urbanism movement to attract a creative class (Florida, 2002). Global marketization efforts have exacerbated competition among cities for scarce resources (Kenny and Zimmerman, 2003), and this has trickled down into neighborhoods (Zukin, 1996). Especially in urban neighborhoods, cultures could be appropriated to please the masses (McClinchey, 2008). Nevertheless, nostalgia and authenticity remain both driving forces and sources of tension in today's neighborhood developments.

A critique is that many cities and communities relying on new urbanism practices seem manufactured or inauthentic (Kenny and Zimmerman, 2003), so nostalgia could serve to rescue that problem. Barnes, Waitt, Gill & Gibson (2006) use Wentworth Street in New South Wales as their case study, talking with project designers, planners, and other stakeholders about a redevelopment project on the street. As they explain, several project planners relied on nostalgia to create a sense of place. Oftentimes, this was an idealized version of the past, using words such as "country town main street," "quaint sleepy," and "old splendor" (Barnes et al., 2006, p. 344). Wentworth Street, though, is quite close to manufacturing and steelworks, so "the 'country' lifestyle they imagine for Wentworth Street never actually existed. Yet this does not prevent the material streetscape of Wentworth Street from being nostalgically re-imagined by this planner as a landscape of faded grandeur and uncomplicated, 'sleepy' pleasance" (Barnes et al., 2006, p. 345).

Similarly, Milwaukee rebranded itself in the 1990s as the Genuine American City (Kenny and Zimmerman, 2004) for its closeness to "real America" (p. 79). Marketers previously wanted The America You Remember as a slogan but such "explicit nostalgia" (ibid, p. 79) could be offensive to segments of the population not included in that idealized vision. The chosen Genuine American City branding was deemed nostalgic enough to attract different stakeholders, including local residents, to Milwaukee. Local residents were being enticed to come back to the city center to take part in public life in a more visible public space than the suburbs (Kenny and Zimmerman, 2004). Slogans and marketing materials focused on Milwaukee's architecture and walkable streets but removes racial struggles present there and in so many cities throughout the US. In that instance, "nostalgia may appear a relatively harmless endeavour when it comes to urban boosterism, but it is not so safe when it informs attempts at urban reconstruction" (ibid, p. 93).

Urban communities are trying to rewrite their own narratives via something called *heritage tourism*. Boyd (2000) details the Bronzeville neighborhood regeneration in Chicago. Instead of succumbing to traditional development and gentrification, residents banded together to embrace and expand their historical roots. Boyd is quick to say this kind of effort could be the exception rather than the rule, but it is worth pursuing to maintain historical relevance in contemporary economic markets.

In the remainder of this chapter, I detail the ways in which interviewees relied on nostalgia to remind themselves of idyllic neighborhood eras. I coded this as a longing for a childhood neighborhood and development as a way to foster nostalgia.

Longing for a childhood neighborhood

When coding the data, I parsed out further the ways in which people used a sense of nostalgia. One of the more popular ways was hearkening to a childhood memory of what a neighborhood was – and consequently what it should be.

For a Lafayette, Louisiana, resident who was born and raised there and is now starting his family there as well, nostalgia crept in when describing his current and former neighborhoods. As he described, it took him several years to find a neighborhood that resembled the feel of where he grew up:

> *When I grew up we had a huge lot in our neighborhood and we spent all of our time outside playing in the neighborhood. We played football, baseball. We rode our bikes. It's a conversation I've had with a lot of my friends. You don't see that as much nowadays in neighborhoods that are being built. . . . I look at our greatest neighborhoods, and this is my opinion, but our greatest neighborhoods are the ones that were built many years ago. The diversity of the architecture, the lot sizes. It's hard to explain. If I'm looking for a house I'm looking in those neighborhoods first.*

When he and his wife moved into their current neighborhood, they did not know anyone there. As older residents began to pass away – in his and other neighborhoods – younger families moved in. As he explained: "For the first two years that I was there, I loved my house. I loved my location, but I wouldn't say that I loved my neighborhood. What happened was that we just didn't know anybody. Then one day, two years into it, we see a young lady walking her daughter around, and I have a young son and we met her. Now all of a sudden it starts to feel more like a neighborhood."

A Realtor in Orlando said he also sees a sense of nostalgia when working with buyers. Home buying, he explained, is mainly emotional rather than purely rational. It is about feeling, and he always makes sure to ask his clients how they feel in a neighborhood. In today's real estate market, he said the emphasis is on community feeling. He named the city of Mount Dora, Florida, as an example. Mount Dora, about an hour northwest of Orlando, is known for its quaint arts community and hosts one of the country's largest outdoor arts festivals each year. The Realtor said people flock to Mount Dora for this arts feel, community events, eclectic dining, and local shopping. (When I visited Mount Dora, I described it as a setting for a Hallmark Channel movie with its small-town charm.) "It was a community that was a throwback to quieter times, to peaceful times. The times that I talked about where I could walk, where I could run around and not worry. My perception was safety. My perception was pride of ownership, pride of community, those sorts of very organic but very emotional comforting feelings." His feelings align with existing research that says walkability and recreation often affect how people perceive a neighborhood and, in turn, themselves (Crompton, 2001).

Communities like these, along with walkability, are what contemporary home-buyers want. According to the Realtor: "People with kids typically are concerned with safety, schools, and traffic because they're worried about the kids and they want to know proximity to things. They want the kids to be able to play like their life. In summertime when school was let out, outside of Washington, DC, in Annandale (Maryland), my mother at 7 o'clock in the morning, she didn't work but I was out of the house playing with my friends. She didn't know where we were, and we were safe having fun. Today that would never happen."

Perhaps this kind of nostalgia ties to what Brown-Sarancino (2004, p. 135) terms *social preservation*: "the culturally motivated choice of certain people, who tend to be highly educated and residentially mobile, to live in the central city or small town in order to live in authentic social space, embodied by the sustained presence of 'original' residents." Social preservationists are separate from gentri-fiers in that the former tries to maintain the integrity of the spaces in which they inhabit to gain an authentic experience. Gentrification tries to influence the eco-nomic future of an area, while social preservation tries to learn from and maintain the past (Brown-Sarancino, 2004).

For social preservationists in the study of various US neighborhoods, the desire for authenticity drove their living choices. In one example, Brown-Sarancino (2004) shares a quote from an interviewee longing for a time when neighbor-hoods were more community oriented, like when she first moved to the area. I found similar information in my data. According to one woman living in Orlan-do's Audubon Park, she chose the neighborhood because it was close to work and tight-knit: "It's a great neighborhood. Everybody is so nice. You used to not have to lock your doors but unfortunately now times being what they are you do." The end of her story is this sense of nostalgia and preservation for the old times.

She explained that people want to live in her neighborhood because it is one of the more established, older communities in Orlando. Audubon Park is walkable to schools, restaurants, and lots of smaller, locally owned shops. Home prices have gone up in the area, which sometimes keeps people out, she said. When asked why she thinks people want to live in Audubon Park, she said: "I think they want it because people, I think, *like nostalgia*. I think people who look to live in this neighborhood always aren't necessarily young people. I think there are older people, more established with older families. The kids aren't infants. They have older kids."

While the resident of Audubon Park aimed to preserve nostalgia in her com-munity, a resident of a neighborhood in the San Fernando Valley in California close to Los Angeles lamented not having what she considers to be a community feel in her neighborhood. "I wish that we were in a neighborhood where the kids moved from house to house and hung out for hours at a time. We don't have a community. The neighborhood down the block has progressive dinners, a book club. They're very tight knit in the way my neighborhood only half a mile away, three-quarters of a mile away is not." While she likes the neighborhood and has stayed for 13 years, she explained that if she had it to do all over again she probably would make a different decision.

Nostalgia is indeed a powerful force when it comes to romanticizing the neighborhood. Kasinitz and Hillyard (1995) studied the Red Hook neighborhood in Brooklyn, dominated by industries that influenced how land is used. The neighborhood appears close to everything, yet bodies of water and highways cut off "Red Hookers" from easily accessing many New York amenities. The authors detail the neighborhood's rise as a dock-working waterfront property to its fall with the influx of industries and factories. They interviewed so-called old timers in the neighborhood, defined not just by age but by tenure living in Red Hook. While old-timers likely experienced the neighborhood changes, the term "old-timer" also could be an inherited title for children who grew up in Red Hook and either stayed or moved away yet kept in contact. Many, if not all, were white, despite the presence of Puerto Rican old-timers in the neighborhood. For these interviewees, their connection to the place represents authenticity (Kasiznitz and Hillyard, 1995).

The old-timers see it as their role to tell the true story of the neighborhood, to not let external perceptions invade (Kasinitz and Hillyard, 1995). "They thus invoke a claim of authenticity based not in the present but the old days. They tell stories of the old days and the area's decline, linking their personal chronology with that of the neighborhood and producing narratives that locate their community and themselves in a context understandable for them and others" (Kasinitz and Hillyard, 1995, p. 152). When pressed, old-timers expressed that some memories of the neighborhood might be based more in myth than fact. That is indeed the power of nostalgia.

Development to foster nostalgia

The second pattern found related to nostalgia was gentrification. Gentrification usually involves economically, socially, and physically replacing one community with another through development policies and practices, whether intentional or unintentional. When gentrification happens, social, cultural, and economic displacement occurs (Ocejo, 2011). Social displacement involves the removal of one group's political capital, while cultural displacement replaces one group's way of life with another's (Ocejo, 2011). Ocejo (2011) examines how a nostalgia narrative played into the gentrification of Manhattan's Lower East Side neighborhood. The people he terms early gentrifiers have a connection to the neighborhood so view creeping establishments such as new bars, restaurants, and nightlife as a threat to their nostalgic claims. Most early gentrifiers have seen the neighborhood thrive, falter, and revive. It is the revival that they sometimes see as inauthentic, paving the way for many a new urbanism development that claims to bring authenticity while threatening neighborhood norms (Ocejo, 2011).

In some of the neighborhoods in smaller cities, people relied on that sense of nostalgia to understand current development patterns – but also push back against them when possible. As the resident in Lafayette, Louisiana, described:

From a historical standpoint, what we are seeing now is two things. Number one, because the city was developed first, as we grow and have grown now, what you're seeing is developments on what used to be the outskirts of town and old farmland. So urban sprawl is happening. That taxes local government because you have to send your services further out in what was farmland. Now there's a big push on developing of the urban core because you have all your services, your streets, your utilities, the land is there. Now it's more sort of a density thing where we are looking to become more dense in the urban core.

He guessed that people who chose to leave Lafayette's core for the more rural areas wanted land, a yard, and quiet. Those coming to the urban core are seeking the mixed-use urban development common in many cities throughout the US, whereby much of the development is mixed-use and people can walk to shops or restaurants. "What a tourist wants in most cases is what a resident wants. You want connectivity. You don't want to have to get into your car all the time to do things; that's whether you're a visitor or a tourist."

Social preservationists will ideally rally against gentrification efforts, such as those occurring naturally or through design (Brown-Sarancino, 2004). For the Audubon Park resident quoted earlier, she lamented the people buy older homes in her neighborhood only to tear them down to build something larger and more modern. Usually the newer houses are two stories, dwarfing neighbors in one-story, ranch-style homes. Her concerns echo those Brown-Sarancino (2004) found, where preservationists worry that development that leads to gentrification also will cause homogeneity.

As explained previously, a trend in real estate is buying older homes to tear them down. In the place of sometimes historic buildings come modern marvels that look out of line with the architecture. There is a sense of loss for some when it comes to this kind of development moving into the neighborhood. As the resident in San Fernando Valley explained: "So you have a rundown old house that's probably owned by an original owner from 1949, and the house next to it looks beautiful and modern and just been redone. No one is buying one of the original houses just to move in. They're buying those to gut down to nothing and rebuild from 900 square feet . . . to a 2,000- 2,100-square-foot home." For her, though, this is not necessarily a negative, as the new development means usually younger families who bring what she called "vitality" to the neighborhood. Perhaps for her, this will bring what she longs for in a neighborhood (emphasis added):

My mother could have left me with any number of parents on the street. I never really felt that with the people here in my neighborhood. I feel like an island here. I know the people. I chat with the people . . . but I never really felt like anybody was there to lift me up.

Here she shares a tension between old and new in the neighborhood. Perhaps as families with children move in, the sense of community will change. She herself was pregnant at the time of the interview so was longing even more for that neighborly feel. And yet, she is hesitant to move because of housing prices and continued sprawl in the Los Angeles area. "We kind of joked after we moved here that this would be the starter house, and 13 years later we look at each other like 'Oh, we are never leaving.' Where would we go? Literally, where would we go?"

One resident of the Forest Hills neighborhood in Little Rock, Arkansas, who is an active member of the homeowners' association and in local politics, explained how development has impacted his neighborhood. First a highway divided the area in the 1970s, and it "destroyed a very active African American neighborhood and business area that was thriving." As the neighborhood began to recover through the years, a new challenge arose: a technological/research park. Homeowners in the area pushed back against the development, fighting against what he said were eminent domain efforts to take properties.

He said people posted "Keep Your Hands Off My American Dream" signs in their yards in protest of development through the neighborhood ("American Dream" as nostalgia). The neighbors were successful in turning back the project, then moved on to supporting the development of new urbanism spaces – mixed-use affordable housing that is walkable. "We are looking at, 'What do we do now?' Everything is turning around, and now we need to brand this wonderful thing that we have going on over here."

For other interviewees, nostalgia through design was a good thing that actually encouraged them to move into their chosen neighborhoods. In one example, a resident of the town of Celebration near Orlando, Florida, said: "We loved the beauty of the town when we came across the bridge. It's like, oh gosh, look at that house. The place was well kept. We didn't know about covenants and deed restrictions and this and that. That's why the place looked the way it did." They liked that the new homes would have new appliances, pipes, paint, etc. The trees, though, were a deal breaker. He said he wanted trees that were close to fully grown so they ended up purchasing a home in the neighborhood a bit farther from the city center (where there are shops, restaurants, and a park). He called the decision "fortuitous" because that downtown-style area of Celebration necessarily receives a lot of traffic. Being a bit farther out allows for easier access to highways yet is still walkable or bikeable into the main neighborhood core. "There's plenty of homes but not too many. There's virtually no traffic. *It's like driving in the countryside in Virginia.*" The last part emphasizes nostalgia in the current neighborhood by thinking of a place he formerly lived.

Ellis (2002) details how nostalgia as a term has long been used to critique new urbanist development efforts. Suburban sprawl came about as automobiles became popular in the United States. All of a sudden, people did not have to live in a central city to access critical services and needs. Cars made it so people could spread out, have more land. New urbanism, though, countered this movement as people wanted more walkability, bikeability, and access to resources without having to take a motorized form of transportation. New urbanism is critiqued for its idyllic ways, claiming that planners "want to return to a fantasy of small-town life, a false past purged of all its unpleasant elements and patterns of domination and exclusion, an illusory world of the imagination" (Ellis, 2002, p. 267). That this nostalgia narrative often is wrapped up within new urbanism is incorrect. Instead, new urbanism takes traditions from the past and modernizes them via technologies and design, such as live-work spaces.

Given the nostalgia trope in the interviews, it is important to consider the historical context of neighborhoods in the US as socio-political institutions. This tension between past and present is certainly not new and should indeed be expected as neighborhoods become increasingly contested spaces. Marston (1988), for instance, examines how an ethnically Irish neighborhood came together to become politically and economically active. Spatial and social organization usually go hand in hand to create either robust or foundering neighborhoods (Marston, 1988). In ethnic enclaves, these forces often collided and, in her study, provided opportunities for robust organizing and political activism within the Irish neighborhood. Voluntary associations were vital to the community's success in the late 1800s and early 1900s, though sometimes these associations distracted attention from other political issues beyond the neighborhood, such as anti-immigrant concerns. Despite that, the associations allowed for a honing of political skills and community solidarity (Marston, 1988).

While Marston (1988) detailed voluntary associations to build a sense of community among Irish immigrants, neighborhood associations remain a strong force in governing processes. There are typically two distinct types of neighborhood associations: mandatory and voluntary. City officials from Orlando detailed the distinction.

Within the city of Orlando, explained a member of the city's CNR team, mandatory associations became popular in the 1970s when condominium complexes sprang up. Mandatory homeowners' associations are when community members must pay a fee to join and live in the neighborhood. Gated communities with homeowners' associations exemplify this concept well. For instance, the city official said that developers who want to put in amenities such as tennis courts, pools, and golf courses, for example, "needed to create a system to support that forever because we are not going to take it on."

Voluntary neighborhood associations, on the other hand, do not compel buy-in so the same level of services might not be available. Regardless of the structure, neighborhood associations are a strong governing force, usually their own legal entities with rules, bylaws, and officers (Logan and Rabrenovic, 1990), as discussed in Chapter 3. Associations also can give people a sense of identity and purpose within the neighborhood (Logan and Rabrenovic, 1990), and can think about development and events that could foster a sense of nostalgia.

Concluding remarks

It is perhaps not surprising that nostalgia finds a place in neighborhood branding and marketing – either implicitly or explicitly. In the US, for example, city branding slogans might refer to a small-town feel within a larger geographic area. Nostalgic thoughts are meant to reframe or even erase past negative experiences, making people yearn for a time that might not have actually existed as they remember. This becomes a powerful force in neighborhoods when gentrification forces come into play and can galvanize people against reconstruction of neighborhood narratives (Boyd, 2000). As Ocecjo (2011, p. 306) explains about the Lower East Side neighborhood in Manhattan:

> *People weave nostalgia narratives in the face of threatened, damaged, or destroyed identities. The nostalgia narrative's power lies in its ability to use selective aspects of the past to construct a new identity that stands up to, defines, and counters change in the present. This identity can then serve as the basis for political action against powerful transformative forces. But nostalgia narratives cannot be taken at face value. People weave them within specific contexts and for specific purposes.*

Nostalgia is a social construction used for specific purposes, thus its power lies in believing the narrative. Interviewees brought up nostalgic thoughts especially related to neighborhoods they remember growing up in. Said one resident of a neighborhood in San Diego who grew up in Pittsburgh, "Growing up in my neighborhood, I lived at the end of a cul-de-sac. We were neighbors in the truest sense. . . . People looked out for each other, helped each other." In his San Diego neighborhood, that was not the case and it was something he expressed missing, saying "I want to see people interacting" similarly to the famed children's show *Mister Rogers' Neighborhood.*

This nostalgic notion comes up in neighborhood design and a focus on safety, a focus on keeping the others out and the desirable in. Nostalgia in place branding can erase critical problems that need changing, choosing instead to present idealized images of the place to increase economic gains (Gotham, 2002). There is no solution to "fix" the nostalgia issue, though one can look into why people use these narratives, how, and when to better understand them. For neighborhood branding and identity, nostalgia is a mechanism to attract certain groups while pining for a so-called simpler time where children could play freely. How these narratives are used becomes part of the marketing and branding strategy for neighborhood, again either intentionally or not.

5 Physical and natural space

I love how peaceful it is. Like, I just love the sense of like calmness and green and beauty. I felt like I was in nature. I just love the beauty, the green of it. My day is so stressful. I'll be out all day. I'm constantly busy, and when I get home I like peacefulness.

– Sabre Springs, San Diego, California, resident

On November 23, 2018, 13 federal agencies issued the Fourth National Climate Assessment warning of severe economic consequences of climate change currently and in the near future (https://nca2018.globalchange.gov/, 2018). The report outlines how human behavior, including transportation and energy production, harms the larger environment. Although climate change is beyond the scope of this text, where people choose to live affects transportation habits, energy consumption, and access to natural resources, specifically green spaces. Population migration patterns historically depend on the physical landscape/geography. Physical features are often incorporated into the larger place branding such as Denver's "Mile-High City" or Minnesota's "Land of 10,000 Lakes." Although human development has negative consequences for the natural environment, the natural environment also can influence development patterns and one's sense of place, neighborhood brand identity, and social connections to neighbors.

This pattern emerged when people were asked questions about why they chose their neighborhood or to describe an ideal neighborhood. I did not ask a direct question about parks or recreation or green space. Many people mentioned features such as trails, green space, walkability, and bikeability when describing an ideal or current neighborhood. A challenge with these features, though, is that they are often "nice to haves" for those in a lower socio-economic status, highlighting the tension between social status and physical health needs when people might not have access to safe recreational spaces (Powell et al., 2006). Green spaces, though, can serve as community centers and physical spaces in which to bring people together to hopefully create a sense of community (Kaczynski, Potwarka, and Sealens, 2008).

One Orlando Realtor explained that today's consumers are more aware of their own carbon footprint and are not only seeking more energy-saving features in homes but also green space around or near them. He explained: "They're more cognizant of green space, walkability, even though our health numbers, if you look at weight gain and stuff like that in adult populations, the reality is it's not going down. . . . On the selection of the houses, people, especially in Florida, they want outdoor living space. They want open floor plans; they want light." Green space helps people cope with stress and should increase other positive health outcomes when access is readily available (Roe, Aspinall, and Ward Thompson, 2017).

In this chapter, I show the connection between people's desires for green space and associated components such as walkability, bikeability, and outdoor recreation. I also detail how neighborhoods and cities can and do market these features as selling points, which could also serve to potentially exclude people from lifestyles they cannot afford.

Branding green space

The exact role that green space plays in public branding campaigns is still emerging in the literature (Gulsrud et al., 2013). Ideally, when used in a branding and marketing sense, the word *green* "encompasses both an environmental policy and biophysical dimension" (Gulsrud et al., 2013, p. 331). The policy side focuses on promoting policies and regulations that encourage sustainability, usually environmental sustainability, while the biophysical side is what most respondents in this research focused on: parks, outdoor recreation, community gardens, green roofs, and walkability. "Thus a green city brand can be related to a vision for (1) increased urban environmental political oversight, (2) an ambition to focus on and develop urban biophysical qualities, or (3) both aspects combined in order to achieve a market advantage" (Gulsrud et al., 2013, p. 331).

Green spaces are becoming a major way cities are attracting tourists (Chan, 2017; Majumdar, Deng, Zhang, and Pierskalla, 2012), using green branding as a means to create an overall image of a green city, attracting people who identify with those same values (Chan, 2017). Chan (2017, p. 192) argues:

> *A city brand embedded in high-quality public green spaces may also embrace elements that are health-related, e.g., pleasant experience of green spaces, trees and landscape beauty, quality and accessibility of parks, etc. These health-related elements may intentionally be highlighted by the city marketers or municipal governments, or unintentionally linked to the city brand that is implicitly valued by the residents or visitors through their experiences in different types of urban green spaces.*

Whether planned or unplanned, green branding can become a powerful tool for attracting various kinds of economic investment either via tourism, new residents, or businesses (Chan, 2017). The desire for green space seems to cross the rural-urban divide, as residents in Lo and Jim's (2012) study of Hong Kong revealed that people, even in a high-density city such as Hong Kong, want access to green space. For some in their study, the desire for green space was pragmatic in that it alleviated some stressors and made respondents forget the hyper-developed confines in which they lived (Lo and Jim, 2012).

Green branding has been found to favorably affect a person's functional and emotional reactions to a brand (Hartmann, Ibanez and Sainz, 2005). Functional elements detail a brand's environmentally friendly policies and practices, while the emotional elements relate to how a person feels about the green branding and positioning (Hartmann et al., 2005). Using corporate advertising as their study design, the authors found that people had both functional and emotional responses though the emotional often proved more powerful. A Catch-22, though, is that a purely emotional campaign leaves out vital facts and figures people also need to make a decision, so Hartmann et al. (2005) recommend a balance.

A problem, though, is that sometimes cities and places will use green branding strategies to only increase financial gains rather than put any sustainable policies or practices into place (Andersson, 2016). McCann (2013) calls this kind of policy *boosterism*, whereby cities tend to push and promote what they deem successful policy interventions. McCann (2013) uses the case study of Vancouver to show how city officials there tout on a global level green policies and practices, ideally trying to have other city mayors promote similar policies back home. Related to high-profile issues such as environmental sustainability, policy boosterism can serve as both peer pressure and a neoliberal economic tool to increase practices that on the surface might seem sustainable but when examined further might be environmentally harmful (McCann, 2013).

Take, for instance, ecotourism, a hot topic given the potentially negative effects on natural resources. While ecotourism strategies might attract more tourists, Bell (2008) found that many local businesses trying to also cash in on the green image had no sustainable practices themselves. The conundrum is thus: "But while tourism produces profits, it incurs environmental costs which degrade the quality of the product. The key problem is that even with eco-tourism, green-ness might not be sustainability implemented; large numbers of backpackers are taken off the beaten track, but it can become beaten into something of a highway. Meantime, however, the green myth is infinitely sustainable: an ongoing project of spin" (Bell, 2008, p. 353).

For some neighborhoods in my study, they embraced environmental features that they could then promote. A resident of the Millers Bay neighborhood in Oshkosh, Wisconsin, explained to me that when he and his family moved to the area, they did not know much about the individual neighborhoods. "The first [house] we looked at was in this neighborhood, and that's the one we ended up getting. The real attraction I would say is geographic above all else. We are about a block and a half from Lake Winnebago so we're really dominated by water, and on the lakefront by our house there's a giant park. It was beautiful to look at, and I would say more than anything else that ability to have that asset of being beautiful and something to use each day." Choosing their house in that neighborhood, he explained, was a bit surprising given they came from Milwaukee where their prior neighborhood was highly dense, "where we were really on top of each other, and we liked it. We were all up in each other's business, but we really liked it. It was a great city neighborhood. We came in thinking that's what we wanted in Oshkosh. . . . It took seeing the natural beauty to make us change our minds."

Alternatively, Audubon Park in Orlando seems to have developed a brand identity as being green and environmentally friendly. This was not necessarily intentional but emerged through time, given some of the neighborhood's walkability and local businesses (and likely its name, referencing environmentalism). One Orlando resident who does not live in the neighborhood described Audubon Park as "rustic, green, it's sustainable. Not a hippie feel but that sort of vibe to it. It's a chilled-out Baldwin Park, I guess." Said another resident who does live there: "We still have all these oak trees. If you want to walk or you want to ride your bike or if you want to take your kids out, you can walk these streets because it's a mature neighborhood. It's a nice community neighborhood to walk in. I think people, that's what they think – it's a nice, quiet, mature neighborhood, which it is."

A desire for green space

It is not surprising that people in this study expressed a desire to be near and enjoy green space. Access to green space has been found to have positive effects on a person's mental health and well-being (Jackson, 2003) though this relationship is not a panacea (Gascon et al., 2015; Lee and Maheswaran, 2011). When combined, exercise and green space exacerbate these positive effects (Pretty, Peacock, Sellens and Griffin, 2005). Frumkin (2003) explains four elements of the built environment that can promote positive health outcomes: contact with nature, building design, safe public spaces, and the urban form itself. Frumkin details pros and cons of each element, asking more questions than providing answers. For example, it is not enough to be near a park but park quality and features matter for health benefits (Frumkin, 2003).

For some interviewees in this research, access to green space was a necessity. Said one resident of St. Petersburg, Florida, about choosing her condo unit: "When we first started looking, I told my Realtor that I wasn't sure what I meant by this but I needed a suitable green space. And I said I don't necessarily mean a lawn. I just needed a quiet, secluded place that was outside. A screened-in lanai, could be a fenced-in backyard. I just needed some place I felt like I could be alone and not aware of my neighbors." A resident of an area just outside of Orlando expressed similar sentiments about wanting space: "As I've gotten older, I work downtown. I deal with people every day, and I deal with all of that so I like that escape now. Where in 30 minutes I'm literally on a dirt road with my little world."

The executive with the Convention and Visitors Bureau in Lafayette, Louisiana, said people, both residents and visitors, are looking for "green space, parks, quality of life." He explained the city has a mix of older and newer neighborhoods, each with distinct feelings. The historic homes, for instance, have a different identity than newer, mixed-use developments. Green space plays a big role because "if you look across the country, some of the biggest tourism-type places like Central Park [in Manhattan] or Chicago where the Bean [sculpture] is, those are things that resident enjoy and tourists enjoy."

I spoke with representatives from the city of Orlando's CNR department, and one program they explained to me is called GreenUp Orlando. Started in 1985, GreenUp's purpose was to encourage residents to plant trees in their neighborhoods to increase not only beautification but also sustainability efforts (City of Orlando, 2019). CNR team members explained that the city provides materials such as the trees and plant material, technical advice, and tools, while "the neighborhood provides the sweat equity," one member said, adding that "the goal is to plant the proper plant in that spot" so city officials will come help with that advice and guidance. The neighborhood becomes responsible for watering the trees, which are purposefully kept on the smaller size so as not to drain water resources from the neighborhood and city. She said for neighborhoods "it gives them a sense of accomplishment" when they undertake projects with GreenUp, given the volunteer time, community building, and aesthetic improvements.

Similarly, an executive with the NoMA business improvement district (BID) also discussed their green marketing campaign. NoMA stands for North of Massachusetts Avenue in Washington, DC. She said the branding of NoMA came about before she took over as an executive in the BID, with the name coming from the city planning office to identify the area that needed redevelopment after the district's near-economic collapse. There was a formal brand-visioning process that involved relevant stakeholder groups, and "the outcome was reasonable. The right answer was finished." In other words, she was happy it was over. "It didn't matter what the name was. People can create those associations." Since that initial exercise, she said the brand has been refreshed several times, focusing on the neighborhood's high-tech offerings and then on City Greener "because the neighborhood is known for being green from the built environment but also from parks."

Green space is part of urban renewal strategies that aim to replace vacant land with outdoor recreation and park spaces to ideally increase economic and social benefits to and for the neighborhood (Schilling and Logan, 2008). A community organizer in Washington, DC, described the importance of these types of non-market spaces for neighborhoods. He defined non-market spaces as those not specifically developed or designed to turn a profit, such as parks, green spaces, schools, community centers, and even cooperative living arrangements. He explained that non-market spaces do not preclude investment. For example, a land trust aimed at keeping housing prices affordable is not prevented from owning retail spaces, and indeed these spaces "are consistently looked at as incubator space or below-market to support social entrepreneurship measures."

Similarly, a Tucson resident and professor at a local university there explained it this way: "Urban development includes a lot of features from the community level. It's ironic that when you establish them in areas that did not have them that the social tissue that was supposed to benefit leaves the area." When gentrification and development does happen, she suggested looking at the local economy to possibly prevent neighborhood flight. "If you say most people do agriculture in the area, maybe they have green roofs and urban harvesting so you can create employment for those people so the increase in prices in community reflects increases in salary."

As more developments spring up throughout the US, either through apartment complexes or mixed-use new urbanism-style development, green space is becoming scarcer (Haaland and van den Bosch, 2015). Urban growth that ignores intentional green space inclusion produces problematic outcomes given the positive relationship between proximity to green space and perceived general health (Maas et al., 2009a). Research also suggests that green space has been linked to lower mortality and morbidity (Maas et al., 2009a; Mitchell and Popham, 2007), decreased stress (Roe et al., 2013), increased social interactions (Maas et al., 2009b; Kemperman and Timmermans, 2014), and improved mental health (Zhang, van Dijk, Tang, and Berg, 2015). The positive effects of green space are mediated by the socio-economic environment, though, so green space and access to trails is often a privilege rather than a guarantee.

Walkability and bikeability

Closely related to a desire for green space was the want for walkable, and relatedly bikeable, neighborhoods. Some neighborhoods use those physical features in their marketing strategies, either purposefully or via word of mouth. For instance, the Lake Nona neighborhood in Orlando promotes its 44 miles of trails on its website. Millers Bay in Oshkosh, Wisconsin, promotes the recreational activities available on and around the lake. People in this research explained a desire for more walkability and bikeability as part of an ideal neighborhood. This makes sense, given the overall health benefits that can come from accessibility to outdoor paths and trails (Diez Roux, 2001). A need to run around, walk to places, bike calmly also has a bit of nostalgia attached, and nostalgia is something people can market and sell (as explained in Chapter 4).

Walkability is a quality-of-life measure found to positively influence someone's place attachment (de Azvedo et al., 2013). In their study of London's High Street district, Ntounis and Kavaratzis (2017) explained place-branding practices are about creating attractiveness, both physically and socially, for the place. This process ideally involves stakeholders along the way, making the process a key governance strategy (Eshuis and Klijn, 2012). Ntounis and Kavaratzis (2017) explain that place branding involves the relationship between myriad factors, including attractiveness (how the place looks), infrastructure (walkability), facilities (retail), and vision and strategy (what the brand means and will do). For neighborhoods, the same logic can apply, especially concerning new urbanism-style developments and branding.

The St. Petersburg, Florida, resident quoted earlier described how walkability was central to her decision to choose a location to live. Her husband, a gentleman in his late 70s, has mobility issues so they wanted a place where he felt safe walking and moving around. They used to live in a large house away from people, and "it was very secluded and peaceful. . . . I lived in a big place that had huge picture windows in each room and only had drapes in two rooms because you didn't need them." They went from that property to a smaller condo overlooking the water, which meant sharing walls with neighbors. Finding their current condo came after looking at many other potential homes. She described one to me as follows:

I loved the space because it was open and it was very large. It was much larger than the place we ended up buying. It had lots of windows, and it had this beautiful light. But my husband looked at it and he kept saying, 'I don't know, I don't know.' And it wasn't because of the price. He just kept saying I don't know. And finally he said, 'I figured out what it is.' He said, 'We're not in a neighborhood.' I said, 'What do you mean?' He said, 'There's no place for me to walk expect the parking lot. . . .' As we continued our quest to find the right place to move to, that was one of our criteria, was it a walkable environment for him? My walk requirements were a little different. I wanted to be able to walk to the beach or to a grocery store or the pharmacy, and I will be able to do that again.

Walkability was also closely linked with nostalgia, feelings of safety, and sense of community in this research. Quality of spaces and destinations is key to creating a safe environment that builds a sense of community, rather than just imbuing a place with new urbanist principles in a vacuum (Wood et al., 2008). Leisurely walking and an ability to see neighbors also shaped sense of community and place attachment (Wood, Frank, and Giles-Corti, 2010), so it is perhaps not surprising that cities and neighborhoods would include these elements in branding and marketing strategies.

As an example, the assistant director of the CNR department in Orlando said walkability often is linked to a feeling of safety and a sense of community. "When you see people out walking around, it's 'Hi how are you?' They're taking care of their yards, so there's pride there. They feel comfortable going next door to borrow an egg." She said a sense of community is heightened when people have nearby access to amenities such as a coffee shop or park because those places tend to be where people congregate. "I think when you see an active neighborhood where there are people out walking, jogging, and kids out playing, that really adds to that sense of community. It feels like, wow, I stepped back in time."

Even when walking is accessible, it does not mean everyone will take advantage. A resident in Millers Bay in Oshkosh, Wisconsin, said that even though the neighborhood lake has paved walking trails leading to the local middle and high schools, most people do not use it. He said a local doctor living in the neighborhood noticed this pattern so decided to help kids get a bit more exercise into their days. "Twice a week we have walk to school with Dr. Eric. He also started a skating program with the kids at school" so when the lake is frozen they can all go ice skate as a community. But for him, the lake and trails are a key reason his family stays in the neighborhood. He said: "I really think it's the water. We can sit in our living room and look at the water, the swans, the ducks, and sailboats and people on the trail all day long. Friends of ours from the west side say, 'Oh my gosh, I've never seen so many people out walking.'

It's just buzzing. We love the activity. We love the view. We always say it's the best place in town. I really believe that."

Concluding remarks

Neighborhoods can brand walkability, bikeability, and green space as part of their overall ethos and sense of place. New urbanism-style developments might have an easier time at this compared with older, more suburban neighborhoods. However, the results of such marketing and branding efforts around walkability related to downtown areas is mixed (Sneed, Runyan, Swinney, and Lim, 2011). Socially vulnerable populations might not have access to sidewalk, trails, or even reliable public transportation, only exacerbating neighborhood-related social justice issues (Bereitschaft, 2017). Nevertheless, these characteristics form part of neighborhood identity and image for people in this study.

The convention and visitors bureau executive in Lafayette, Louisiana, explained to me that they use that connection and walkability in their branding campaign. He said the Centers for Disease Control and Prevention named Lafayette the happiest city in America in 2014, and since then they have expanded on that brand identity. From a tourism standpoint, he said they still run ads under the "Stay Happy" campaign umbrella featuring local, Grammy Award–winning musicians. "That's really how we are every day, and so when I think of a neighborhood, if you don't have that in a neighborhood, if you don't have people waving and smiling, stopping to talk to you while they walk their dog, that's not a neighborhood I want to live in."

A commercial architect from Dallas, Texas, told me that a desire for walkable, bikeable spaces has changed her business, especially when it comes to landscape architectural elements. For example, when we spoke she was working on a corporate project for an energy company relocating to San Antonio. They were working on a space along the city's famed Riverwalk, an open, outdoor, commercial space that is pedestrian friendly. She said the building's patio would be deemed public space because it opens right onto the Riverwalk. She said the design team also was looking at putting a bikeshare on site for not only employees to use but also those who pass by the building while along the river. "It's harder with corporate headquarters because especially where you are and getting into the public versus private and understanding the fine line of that, but the idea of engaging communities on a level of design is very important."

6 Neighborhood identity and "Like Me" politics

On [the] other side of stadium and tracks, homeless people or lower income . . .
they loiter, and they'll come to our neighborhood and just sit on stoops and sit
outside restaurants. . . . The crime is whatever the crime is, but I'm not saying it's
necessarily the homeless populations specifically, but it is an issue.
— Baltimore, Maryland, resident

When analyzing the data, I noticed a pattern where people used coded language to explain why they chose a certain neighborhood. This is perhaps not surprising given a "like me" bias (also called affinity bias) forms when we feel a connection to someone because, say, we went to the same college or live in the same place (Oberai and Anand, 2018). Within the neighborhood setting, Quillian and Prager (2001) found that neighborhoods with more young black men are perceived as dangerous compared with whiter neighborhoods. This perceived bias not only affects the neighborhood residents but also the neighborhood's image and can brand it as unsafe given these stereotypes. Negative stereotypes and geographic clustering of people into economically depressed neighborhoods does have long-term effects on poverty rates, especially for black residents (Quillian, 2003).

In this chapter, I share some thoughts from interviewees when it comes to potential unconscious biases in neighborhood selection. I present this without judgment, without making claims about racism or sexism. This was the language used when people wanted to explain why they chose a neighborhood where people looked like them – and avoided neighborhoods where people do not. Indeed, a neighborhood's brand and image – and even strategic branding strategies – play into this perception.

Neighborhood branding and perceived image

In the data, I found that professionals such as economic development experts, city officials, and community organizers usually had a favorable view of neighborhood branding, touting the positive economic advances and return on investment. Most residents to whom I spoke, who were not professionals in this area, were more ambivalent about neighborhood branding. This divide is quite interesting, given literature tells us branding is a strategic governance tool meant to build community (Eshuis and Klijn, 2012). One community development professional, though, took a more negative, nuanced view toward neighborhood branding, so I start here with his thoughts.

"I'm not convinced I've ever sat down with somebody who sat down and explained why. Part of my instinct says don't do it," he said when I asked him why a neighborhood needs a brand. The expert lives in Washington, DC, and has spent decades in community development and community organizing. He runs an organization in DC that focuses on gentrification and its effects on people and places. He explained his time in college helped him try to find ways to bridge the gap between community development and community organizing because "I actually think in many respects currently they are very separate, and I think in ways that is a detriment to both."

He said many community development efforts in the US started in the 1960s and '70s as a grassroots response to a practice called *redlining*. Redlining was a policy of the Federal Housing Administration in the 1930s to expressly deny mortgages and mortgage insurance to black families (Rothstein, 2017). Rothstein (2017) argues that this historical economic oppression carries through to present day, whereby black residents pushed into public housing were not afforded the same educational opportunities as white counterparts, thus continuing the cycle of oppression and depression. These kinds of practices – redlining, white flight, self-segregation – are a result of systemic rights violations rather than one-off policies. Redlining is but one example of how the housing market supports segregation. This book is not about this topic specifically, but this chapter is included because people still use coded language and implicit bias when selecting a neighborhood.

Returning to the community organizing expert in DC, he said that community development efforts in the 1960s and '70s were a response to more equitable (at least on paper) lending practices and grassroots efforts to improve communities that faced unrest during the civil rights movement as a mechanism to return investment into those areas. "Over time the calculus shifted and instead of it being a grassroots-led effort it was a technocratic effort" that severed the community-organizing portion from the economically driven development portion. When economic forces take over, he said, questions of culture and history are often pushed aside, leading to gentrification. "When the market starts working, it's going to work for the people it's intended to work for."

When he and his team go into a neighborhood, he said their task is to question the underlying values embedded within a market system. "We ask questions about what is development bringing that we want, but what is development bringing that we don't want. We ask what we want to keep and what we want to change. So those questions help to create the framework for them saying, 'Okay, maybe not all development is good development, but maybe we need [to] develop along certain avenues.'" For example, he said perhaps a community does not need fancy sit-down restaurants or grocery chains that cater to a wealthier population but needs instead small businesses that bring chances for jobs for local residents.

"I think it is worth having conversations that there is more to our communities and our lives than just economics. So what are the non-market things in our world that we want to protect and create space for?"

He and his team have done work in the Shaw and U Street neighborhoods in DC, both hit hard by gentrification while still working to maintain this sense of authenticity and connection to their black roots. Gringlas (2017), writing for National Public Radio (NPR), told the story of longtime Shaw resident Ernest Peterson, who said he feels like a stranger in the neighborhood he has called home for more than 40 years. "'I go outside, and these people who been here for 15 minutes look at me like, 'Why you here?' That's that sense of privilege they bring wherever they go," he said in his front yard on a sunny Saturday in November. "I been here since '78. They been here six months or a year, and they question my purpose for being here'" (Gringlas, 2017, p. 3). Gringlas (2017) explains that many residents in lower-economic communities feel as if gentrification and redevelopment will not happen to them because of a traditionally strong (though potentially negative) neighborhood image. The community development expert expressed a similar take: "I think if you go back to the '80s and you ask somebody will this neighborhood ever have to worry about that (gentrification), they'd go, 'Are you crazy?' You look at somewhere like Flint, Michigan, somewhere like Baltimore, you might say they won't gentrify" but could similarly to those DC neighborhoods where he has worked.

As Wherry (2011) explains, the market forces that create gentrification also try to strike a balance via cultural preservation and authenticity. The problem, though, is that authenticity is then also commodified and distorted. Said the community organizing expert:

> In DC, the culture of a community, whether it's in a historically Latino community or like in DC with a large jazz history, that becomes part of what makes the neighborhood sexy. They want this sense of authenticity or we are living in a hip neighborhood, but we don't want the black or brown people who made that history present. It's a double whammy of we are taking over somebody else's community and to some extent with doing it, attracting the people who will displace them. With some of the neighborhood branding and marketing piece, it invites the commodification of peoples' cultural identity that ultimately gets used to flip the neighborhood.

He gave the example of Chinatown in DC. He said the area did not normally have a heavy residential presence, and that small presence is all but gone as national and international corporations built up the area to resemble Times Square in New York City. Zheng (2017) wrote that Chinatown emerged as an ethnic enclave as fear and anti-Chinese rhetoric spiked after the 1872 Chinese Exclusion Act. DC's original Chinatown emerged in the 1830s and was thriving until development of the Federal Triangle area forced many in the neighborhood to relocate to a new Chinatown area in the 1930s (Zheng, 2017). "Today, DC's Chinatown is only a distant memory of the once-bustling enclave of the past. Plans to improve commercial development in the area continue to come at the expense of heritage. Gentrification has further driven up rent prices beyond the reach of the majority of Chinese residents" (Zheng, 2017, p. 6).

Said the community organizing expert: "This is what our Chinatown has been reduced to. I think this kind of stuff ends up happening, and that's one of my fears and discomforts with neighborhood marketing and neighborhood branding is what it's intended to do." He recognizes a "solution" to gentrification is not easy so encourages communities at the grassroots level to ask questions and seek longer-term solutions. He shares an example of a community that is a food desert, with little to no access to fresh foods. A typical solution is to bring in an expensive grocery store that pads the company's bottom line and often offers food local residents cannot afford. Instead, a solution would be a mom-and-pop store that can bring jobs and affordable food to the community, giving people sustainable skills rather than perpetuating inequalities.

Branding and gentrification

The foregoing story from the community organizing professional shows the potential problems with neighborhood branding and marketing, especially when they are tools to encourage (purposefully or otherwise) neighborhood gentrification. I spoke with several people who expressed an internal struggle with gentrification – they know it can have negative consequences, can critically think about those, yet still want neighborhoods that conform to an unconscious bias. Unconscious bias and implicit bias are nearly synonymous, both referring to how past socialization practices unknowingly or unwittingly steep into our present-day thoughts and actions (Quillian, 2008).

I spoke to a man who lives in the East Riverside neighborhood in Austin, Texas. He moved there to be with his girlfriend, though "my mother lived here in the '70s, and it was the ghetto I guess. This was a Hispanic neighborhood and it still is in some respects, but it is gentrifying more and more." Ghetto areas typically refer to neighborhoods with a poverty rate of at least 40 percent (Wilson, 1991). Tied closely to this physical segregation are the banking practices noted earlier such as redlining (Rothstein, 2017) and the exodus of wealthier minorities from ghetto neighborhoods, taking with them political, social, and economic capital (Wilson, 1991). Though the term *ghetto* has an official definition from the US Census Bureau regarding the 40 percent poverty rate, the term is usually a shorthand to mean where minorities and criminals live (Domonske, 2014). Domonske (2014) tries to trace the origins of the word "ghetto", beginning with an enclosed area where Jewish people typically lived (or better yet, were put). The word became more known during the Holocaust era, with Jewish ghettos emerging throughout Europe (Domonske, 2014). "Jewish ghettos were finally abolished after the end of World War II. But the word lived on, redefined as a poor, urban black community" (Domonske, 2014, p. 6). The word "ghetto" allows for othering and an easy label to place people into boxes (Coates, 2009). Domonske (2014) demonstrates that as the word made its way into popular culture, especially in hip-hop songs that celebrated the ghetto lifestyle, it lost its mooring to forced segregation and instead became a life choice – a potentially disastrous rhetorical shift with practical effects.

Another person in Texas, who lives in the Farmers Branch neighborhood in Dallas, said he also moved there to be close to his girlfriend. They decided to buy a house after living in an apartment and tiring of loud neighbors, and the unit was "in a bad part of town." The house they chose is in a mostly Hispanic neighborhood. Though there is not a lot of nightlife and the "couple nearby bars are dive bars" they like the location because it is close to the tollways. "It's a combination of location and price. We got a two-bedroom house with a garage for $1,400 per month. It's huge. It's real nice. So moneywise it falls right in our budget. The location for getting around is great. Just because it's a lower-income area doesn't meant that it's bad. It's not like a ghetto, but [you] can tell some people are not well off."

Morrill (1965) notes that ghetto neighborhoods are so pervasive that all large urban areas in the US have them. He notes that neighborhood identity and personal identity are intertwined, putting it thusly (1965, p. 339):

> *Inferiority in almost every conceivable material respect is the mark of the ghetto. But also, to the minority person, the ghetto implies a rejection, a stamp of inferiority, which stifles ambition and initiative. The very fact of residential segregation reinforces other forms of discrimination by preventing the normal contacts through which prejudice may be gradually overcome. Yet because the home and the neighborhood are so personal and intimate, housing will be the last and most difficult step in the struggle for equal rights.*

Ghettos are about othering. People have a bias – either explicit or implicit – toward living with people who look like them and act like them. This manifests in race (Sampson and Sharkey, 2008) but also in terms of health and wellness (Bilger and Carrieri, 2011), bicycling behavior (Pinjari et al., 2008), teenage dropout and pregnancy rates (Harding, 2003), and children's achievements in basic cognitive development (Sastry and Pebley, 2010), to name a few. We gravitate toward where we feel familiarity.

An executive for a developer in Orlando, who was born and raised in the city, described his work with community building and place making. He works for a company called Tavistock, which built neighborhoods in Orlando, including Lake Nona and Isleworth Golf and Country Club. He said Tavistock bought Isleworth on the steps of a courthouse in 1992 as it was going through foreclosure proceedings. He said their first steps were to clean up environmental concerns, build a country club, and make the homes less dense (anti-new urbanism). When the company purchased the neighborhood in 1982, the Orlando Magic basketball team was just coming online. "If you were coming into new money, where were you going to live? The places you could live in Orlando were racist places, at least in the 1980s. We embraced new money. We embraced people of demographic differences. All this money was coming into Orlando, and this was a place where they could be because we aren't a bunch of racists." Isleworth is home to an Arnold Palmer–designed golf course and has large estates where celebrities like basketball player Shaquille O'Neal and golfer Tiger Woods have lived.

In another story of redevelopment, an executive with the NoMa business improvement district (BID) in Washington, DC, explained the history of her area. NoMa (North of Massachusetts Avenue) was a former industrial area with very few residents because its location near the railroad lines, once a boon, became a burden as the highway system and trucking expanded. The neighborhood, she explained, was once home to Irish immigrants in the 1900s but slowly faded to make way for the railroad. "In 1990s when DC was bankrupt, there was an effort to figure out how to improve the economic prospects of the city. One of the things the plans focused on was taking advantage of the Capitol-adjacent, transportation-adjacent areas. That's when the neighborhood needed kind of a name. They needed something to call this conglomeration of vacant properties for the most part. That's how NoMa got coined."

She said the NoMa name was used in the DC planning office but really came to fruition as the BID started promoting the brand. The BID began in 2007, she said, but the name was in use for many years before that. "At that time and until like 2011 or 2012, people were like what the hell is NoMa? Because, not surprisingly, affordable housing and subsidized housing was placed on the perimeter of the city largely. The people who lived in NoMa or NoMa adjacent were people who had been economically marginalized so they felt this whole thing was really insulting generally." She said people resisted the name, feeling it was now a symbol of gentrification, but the problem was there was no solid identity there before, paving the way for a new branding campaign.

Even though the NoMa name did not come from the community, the BID still took steps to promote and brand what they felt was authentic to the area – technology, transit, and wireless connectivity. She said they communicate the brand through various channels, including print and digital media, neighborhood giveaways, and other materials. When asked to describe the NoMa brand she said, "connected, fiber optic wires in the streets, free WiFi in the neighborhood, connections to Congress and government agencies." She used the word "authentic" several times, so I asked what that means: "It's like pornography, you know when you see it. If you're making up a bunch of stuff it's not authentic." The reference is to the US Supreme Court case *Jacobellis v. Ohio* when Justice Potter Stewart commented that he knows hard-core pornography when he sees it. (The case was about obscenity related to a neighborhood theater showing a film some deemed inappropriate. The theater owner was fined by the city, then sued.)

Relating to the NoMa neighborhood, the BID executive explained they have a lower crime rate compared with other areas around the Capitol, so use that as a sales pitch. As a story in *The Washington Post* describes, NoMa is no longer "the wrong side of the tracks," according to the headline (Hoffer, 2015). They attract wealthier people and, by nature, have displaced others through the development. The executive said: "There's a relationship between the capital assets in a neighborhood and the reputation of the neighborhood quite often. Not always.

The typical useful life of a building will be 30–40 years. There's a cycle of investment and disinvestment."

The news headline referring to the "wrong side of the tracks" alludes to racial segregation practices that often saw railroads divide communities, creating a so-called wrong side of the tracks. As Ananat (2011) explains, railroad tracks were visible markers that delineated communities and could be purposefully placed, especially compared with geographic markers such as rivers or even roadways. Jacobs (1961) details the effects of borders, including railroad tracks, noting that "a district lying to one side may do better or worse than the district lying to the other side. But the places that do the worst of all, physically, are typically the zones directly beside the track, on both sides" (p. 257–258). She explains the railroad tracks, waterfronts, campuses, and more borders often serve as the physical and psychological end – or stopping point – of a neighborhood and of its use value. People do not see the value in developing or enriching these places near borders. "The more infertile the simplified territory becomes for economic enterprises, the still fewer the users, and the still more infertile the territory. A kind of unbuilding, or running-down process is set in motion" (p. 259).

Lived experiences

It is not my task in this book to delve into the deep literature on neighborhood gentrification. Instead what I noticed was people using coded language to express a desire to live in a place where people mirrored them in terms of race, socioeconomic status, or commitment to wellness, for example. Parents also cited wanting neighborhoods with good schools, which usually meant predominantly white neighborhoods. I heard stories of people feeling badly for living in neighborhoods that matched their desires but they knew had displaced other populations.

As several professionals pointed out, gentrification is a tricky issue. Said the executive in a Convention and Visitors Bureau in Lafayette, Louisiana:

> *What we think is great maybe some people don't think it's great for the neighborhood so it forces people out. It's a balancing act to those conversations. I don't necessarily believe that argument myself, but I'm also not in somebody's shoes. If my house is worth three times what it used to be, how is that a bad thing? Rent is trickier because you could be forced out. It's a complex discussion. I think people would probably not talk about it. It's hard to achieve what needs to be done when you factor in the possibility of gentrification. I don't know of a community that's got it right.*

There is a link between neighborhood branding and gentrification, as Keatinge and Martin (2015) note the consumption-based practices often associated with both branding and gentrification further exacerbate social class differences and access to communities. There is cognitive dissonance between branding a neighborhood as edgy, hip, or cool while also trying to present the "sanitization of urban locales" usually in the scope of new urbanism (ibid, p. 869). While neighborhoods might have physical barriers that keep people out, such as gated communities, brands, while not tangible, can serve the same othering effect by removing people who do not fit within the brand ideals (Keatinge and Martin, 2015).

Zukin (1987) explains that gentrifiers, perhaps paradoxically, brought attention back to the city and away from the suburbs. In this manner, gentrifiers were seen as different than other middle-class people; they had certain education levels, likes and dislikes, and access to resources gentrification afforded them not found in the suburbs (Zukin, 1987). There also is tension when middle-class gentrifiers enter a neighborhood, having often-conflicting views with locals and longtime residents who might be displaced by efforts of historic preservation. Interestingly, others forsake place attachment for an increase in personal capital investment (Zukin, 1987). Karsten (2014) also highlights similar tensions between the often displacing effects of gentrification with the possibilities of increased economic opportunities for existing neighborhood residents.

One of the interviewees described this tension. He has lived in and around Orlando since his family moved to the area when he was a child. He recently retired from public service after more than 35 years working for the city government. He has seen many Orlando neighborhoods change, especially those surrounding the downtown core. He pointed specifically to the basketball arena and soccer stadium where the Orlando City Soccer Club plays. Both are downtown next to several historically black neighborhoods that have lower economic standing. "Downtown is getting to a point where it's not as safe anymore. Parramore is a historically high-crime area in Orlando. It has a lower economic base. That's why we built the new arena in the Parramore district, because the property values are low." Additionally, the University of Central Florida opened a campus downtown near Parramore in 2019, but "the downside is you're pushing out communities that are there. . . . Families are being displaced at property value, but they can't find anything for that today."

He explained that gentrification dynamics are often hard to control, given city officials and homeowners have no control over market influences. And yet, he continued, there are some externalities that come from gentrification such as new businesses and potential access to jobs (see also Zukin, 1987). "Every time you change something, something's going to go. Like with the soccer stadium, now you've got new businesses opening and things opening that would have never opened there before. But that's either going to bring in different people because now they can say this housing is cheap so they can tear it down and put in another house. You would have never had that if it wasn't for that stadium."

Gentrification often pits people in an us-against-them duality (Keatinge and Martin, 2015). One of the interviewees described in detail how this othering affects her on a daily basis. She lives in a neighborhood in Orlando called Baldwin Park. Baldwin Park is a new urbanist development that sits on what was an old US naval training center. There is a mix of million-dollar homes and more affordable apartment complexes. Baldwin Park has a reputation for being a wealthier neighborhood that excludes those who cannot afford the price on entry. (As you may remember from the introduction, I live in this neighborhood. I live in a one-bedroom apartment; however, I still face odd looks when I tell people I live in Baldwin Park.)

The interviewee moved into Baldwin Park after becoming pregnant with her first child. She openly identified as a Latina female who originally moved into an Orlando neighborhood where there was a larger Spanish-speaking population. Where she and her husband originally moved was closer to the city's tourist district so nothing was really walkable either. During their original house hunting, they had looked at Baldwin Park but dismissed it because "we were turned off by its demographic. It didn't seem as diverse to me. It's a very white neighborhood." When she became pregnant, her priorities shifted.

I wanted a place to take the baby in the stroller and walk around in a safe space with good schools if we did decide to stay long term. It's a lot more family oriented. I can go out and there's other mothers with babies. There's a lot of people walking and exercising by the lake. I like that aspect of it. It feels more like a small town rather than this huge dispersed space where you don't get any personalized feel.

Living in the neighborhood, though, comes with challenges for both her and her husband, who is Pakistani and a doctor at a local hospital. She has been racially profiled in the past, told to speak English when she was doing doctoral research fieldwork in Spanish. "It reminds you that this is still very much the South, and I think that's something that people forget about Orlando. Talking about identities, people don't associate it with the South. It's associated with the Mickey Mouse mentality and tourism, but people forget that racism is still very much alive, and there's still very much this kind of confederate mentality. I mean, we see confederate flags here." She told me of a time walking in Baldwin Park that she forgot her phone in the stroller, hoping nobody would call the police on her for looking suspicious walking among the larger homes.

She was raised in the Bronx in New York City in a largely Latino neighborhood. "Spanish was the language you heard spoken on the street." She described a tight community with "grandmas looking out the window yelling at you to come in." (This parallels what Jacobs [1961] described in her research.) She said she was not sure if she experienced racism while growing up in the Bronx, but she became acutely aware of discrimination when she went away to college at Villanova University in Pennsylvania. "I was very much in a bubble when I was in New York," she said, noting that Villanova highlighted differences for her in socio-economic, political, and social aspects.

When she lived in her former Orlando neighborhood near the tourism center, she would tell people where she lived and they would react by saying, "Oh, there's a lot of white people there. There was this perception, and it was projected onto me." That projected image did not improve when they moved into Baldwin Park. Given that her partner is a doctor, "we have access in a monetary sense to certain spaces that other people might not." Baldwin Park is relatively low in crime, which attracted her to the area as a new mother. Yet while she is not often afraid of the threat of physical harm, "there's danger for me as a person of color in other ways." She described how she has learned to navigate predominately white spaces, being aware of how people perceive both she and her husband in those spaces. But living in Baldwin Park means more opportunities for her son, and "I rationalize that in my head as, 'Okay, this is more important in this moment.'"

Concluding remarks

An economic development professional based in Washington, DC, said gentrification, branding, and marketing often go hand in hand. He said his group works with neighborhoods "that used to be cool and want their mojo back." In his work, he encourages neighborhoods to explicitly state the goals they want to achieve thanks to a concerted branding and marketing strategy. General goals such as increasing economic development are not as concrete as increasing the number of small businesses or improving walkability. He explained:

> *I think almost every single project that I work on, gentrification is an important element, and it's a real thing but it's such a complicated thing. It's like, in my mind everybody has a different definition of it too, but the idea of someone being pushed out that can't afford to be someplace any longer is a sad thing. But it's part of a really complicated set of conditions of which many of them are good, some of them are bad, and the mix has little ability to interject morality and ethics in it. Every single project someone asks about gentrification. If there was a quick, easy answer to that, I would be famous for having written it and also just there can't be.*

He said he encourages people to avoid saying they are worried about gentrification in a blanket manner but to focus on what specific things they are worried will leave or change the neighborhood. For instance, is a community worried about commercial gentrification? Residential? Types of businesses coming in? Health disparities? "Instead of harping on general gentrification, what specifically are the things you'd find unfair or inequitable, and what's the right way to address those specific issues?"

Gentrification projects might see the neighborhood transform from a geographic place to one that is mental and emotional (Ulldemolins, 2014). This shift can turn a neighborhood from a place to live into a complex association of consumptive elements. "Thus, the branding process may generate gentrification processes that distort the location's unique and authentic character and transform the location into a standardized space" (Ulldemolins, 2014, p. 3030). Peck (2005) further shows the effects of the Creative Class mentality on neighborhoods and cities, detailing how much of Richard Florida's Creative Class logic actually further exacerbates differences between the so-called haves and have nots. Focusing on the hip and cool, Peck (2005) explains, ironically creates neighborhood seeking homogeneity rather than heterogeneity. Isomorphism is in place when neighborhoods try to mirror seemingly successful patterns within examining what makes their places unique.

This chapter highlighted both how interviewees used coded language to appreciate effects of gentrification while also being critical of its practices. It also showed people dealing daily with the effects of gentrification. Neighborhood branding strategies, importantly for this book, could exacerbate gentrification's problems and positives. There is a delicate balance to strike, and no solution is perfect. Either way, the key is to remember that real people are affected by gentrification practices – and keeping them in mind is imperative.

7 Conclusion

As a community leader, I would be interested in what the perception of my community is. In other words, how do they view my community, and is it what I want them to perceive? And then if it's not what I want them to perceive, how do I get that image out there?

– Celebration, Florida, resident

This book is a step toward developing working definitions of a neighborhood brand and branding. For public administration scholars and practitioners, this is meaningful because more public entities are spending real dollars on strategic branding projects that are becoming key governance strategies (Eshuis and Klijn, 2012). It is no surprise that branding practices that seem to provide economic returns at the national, state, and city level are trickling down into neighborhoods. Neighborhoods are where people spend the bulk of their time and are where they make one of the biggest purchasing decisions of their lives, so more research is needed into the importance (or not) of brands and identity.

Throughout this book, I detailed elements people thought were an important part of what constitutes a neighborhood brand: sense of community, feelings of safety, connectivity and engagement, schools, green space, walkability and bikeability, and nostalgia. Some of these had coded language involved that people used to mentally separate the "good" from the "bad" neighborhoods. Return on investment was usually measured in economic gains (more residents, more business owners, etc.) but social, cultural, political, and network returns are vital as well. As such, I offer the following working definition of neighborhood brands and branding based on these results:

A neighborhood brand is the combination of identity (what neighborhood leaders might actively communicate) and image (what people think of the neighborhood) that serves as a cognitive shortcut for economic and social decisions (such as opening a business, renting or buying a home, playing in the parks, etc.), and a mechanism to create emotional (either positive or negative) connections.

Neighborhood branding is an active, co-creative process involving neighborhood stakeholders setting mutually agreed-upon goals and values to achieve outputs and outcomes that could include pragmatic and emotional appeals.

The first definition takes into account the realization that your neighborhood already has a brand image in people's minds, so trying to shape the identity through an active branding process can potentially counter negative brand images or shore up positive ones. The resulting neighborhood brand should match reality within the neighborhood, and marketing pieces can be used to communicate the outcomes and outputs. Pragmatic reasons for choosing a neighborhood could include walkability and bikeability, good schools, along with green space and access to resources (such as a grocery store, close drive to work, etc.), while emotional reasons include sense of community, feelings of safety, and nostalgia. This makes sense given brands are emotional appeals that serve as cognitive shortcuts when it comes to decision making, and those decisions are often practical in nature. (Where do I take a vacation? Which detergent do I buy? What brand of cellphone do I want?)

I use outputs and outcomes both because they are conceptually different. Outputs are the results of the process, whereas outcomes are things that take place because of the outputs. Outputs of a branding process could be, for instance, a strategic branding plan, a new logo, community events. Outcomes could be neighborhood pride, sustained leadership, increased place attachment (however the community decides to measure success). For example, in Baldwin Park where I live, there were only a handful of coordinated community events, which was a negative part of the neighborhood identity. A neighborhood events director was hired, and now there are more planned community happenings on the horizon. Outputs and outcomes.

It should be noted these are not catch-all definitions, and the study is not meant to be generalizable to an entire population given its exploratory nature. Instead, these are elements people search for or say are important for a neighborhood, and city officials can begin incorporating more of these into their outreach efforts while neighborhood leaders also can do the same. I should note that these do not happen in a vacuum and sometimes cannot even be created. As Byrne (2013) articulates, neighborhoods provide spaces where you can meet up with people and form social bonds that serve as a foil to hectic workdays. "Such a neighborhood conveys an indigenous identity created by the efforts of diverse people over time, rather than marketing an image deliberatively contrived to control the perceptions of customers" (Byrne, 2013, p. 1597). This might be true, that a neighborhood brand and identity bubbles up naturally, though that does not mean that like city branding or nation branding some proactive efforts and steps cannot be taken. Indeed they are, as the examples from my data indicate.

Example 1: Madisonville, Cincinnati, Ohio

A representative from the Madisonville Community Urban Redevelopment Corporation in Madisonville, a neighborhood in Cincinnati, Ohio, explained that his organization uses social media to reach a wide audience. He said sometimes there is a divide between Madisonville and the surrounding, usually wealthier neighborhoods. Madisonville likes to trade on its grittier, working-class nature, and nostalgia. His organization began working in 2013 with a group called Place Matters to develop and rebrand the neighborhood, along with several others surrounding Madisonville.

He said the redevelopment corporation received a small grant through Place Matters to develop a brand so began working with a local commercial development real estate company and other neighborhood stakeholders to undertake a strategic branding process, beginning with visioning. Unfortunately, though, he said that traditional process did not quite work. "We didn't quite figure out who we were. I don't know that we ever have." He said they then turned to the question of how to make the neighborhood interesting to people, to overcome its brand image of unsafe and rundown. The biggest competition was a development in Cincinnati called Over-the-Rhine, once named the most dangerous neighborhood in the US (Woodard, 2016). As Woodard explains (2016), the redevelopment was led by the Cincinnati Center Development Corporation rather than the local government, and locals have mixed views on the gentrification taking place within the neighborhood boundaries. Woodard's (2016) reporting deftly details the changes to Over-the-Rhine, and I encourage reading the full text. But for the Madisonville representative to whom I spoke, their challenge was to market as the anti-Over-the-Rhine. "They're edgy. We're not. . . . Come to Madisonville, we're edgy without being too edgy and you can have a yard. . . . It's really hard to say we're stable, we're good drivers, and we're safe."

Though the initial branding process got off to a slow start, he said they tried again and contracted with a local firm in Cincinnati to do the work. He said the first step was a focus group made up of diverse neighborhood stakeholders, including residents, business owners, and others active within the neighborhood. Sometimes people showed up, sometimes not. "I get it. If you're a neighborhood stakeholder and you have strong opinions, you're probably really busy." After the focus group came a town hall meeting, which also included a suggestion box for ideas that were not shared during the discussion. He said they had about 60 people participate because "we wanted to find six groups of people that we really wanted to get a good cross section of the neighborhood there. And we managed to do that."

Those ideas went back to the marketing firm to develop a visual brand identity, which was then workshopped with the community members, he said. In addition to a logo, the small-grant stipulations also required physical neighborhood changes (outputs and outcomes), including public art, signage, and events. The logo is a colorful, dynamic letter M and the tagline is Soul of the City. He said the brand rollout has been slow, as they are looking at other ideas such as adding additional branded signage for local business, putting the logo on giveaway items such as tote bags, and hanging banners on construction fencing. So how have people received the brand? "It's been good. As with everything else, there are always going to be some people who don't like it."

Example 2: Lake Nona, Orlando, Florida

Whereas Madisonville representatives hired a firm to develop a brand, Lake Nona's began with the community planning process. Lake Nona is a relatively new community in Orlando known as the home of Medical City for its various hospitals and other medically linked corporations. Lake Nona promotes its health and well-being, along with its high-tech amenities. An executive with the Tavistock Corporation that developed Lake Nona gave me some background, focusing mainly on a neighborhood called Laureate Park in the heart of the Lake Nona area. He used to be in charge of the Lake Nona branding and marketing but recently shifted into other commercial development roles within the company.

He said Lake Nona is interesting because there are so many different parts to the community – a country club, single-family homes, apartments, and townhouses. "Nothing is entry level here just by the location and the economics of trying to make a real estate deal work," he said of the housing prices. He said there are market-rate apartments and townhomes around $200,000 all the way up to mansions in the country club. "For us it's a wide net not only in pricing type but project type," which is important to keep the area thriving in case of economic downturns. For example, during the housing market downturn in 2010, Lake Nona was still developing thanks to the upcoming hospitals and health-related business endeavors. In the Laureate Park neighborhood, he explained that the median home price is about $500,000 and nearly all the residents have a college degree of some kind, and about 45 percent of residents have an advanced degree.

He said when Lake Nona was developing during 2007 and 2008, they embraced the medical city branding. "Given all the publicity and what was happening here, I don't know how you could not lean into that because it's like why wouldn't you?" Though he admitted the medical city moniker was not all that sexy or appealing, it was and still is the reality of the place. As some of the medical facilities were slower to be built, or fizzled out totally, they shifted their branding to performance districts for health and training. That shift resulted in attracting the United States Tennis Association's training facility, along with the worldwide training headquarters for KPMG, a global financial services company. Lake Nona is "such a large project there isn't one theme for everything. It depends who you ask and what the motives are."

A bigger marketing blitz came as Lake Nona continued to grow. He said they did something called Lake Nona-ology, educational pieces meant to tell people things they did not know about Lake Nona. "It was a play on there is so much here and so much to learn about here it needs its own subject." They also did advertisements on buses, billboards, and digital communication. One difficulty they had to overcome early on – and still have to overcome to some extent today – was their geographic location. For many who have lived in Orlando for a while, Lake Nona seems far away. For those new to the area, it does not feel so far given the quick access to highways. Either way, he said the point for their marketing now is to deliver on proof points, or what could be called the authentic elements of the neighborhood. "It became less of 'trust us' to 'hey, it's happening.'" There are many community-run events in Lake Nona, including food truck nights, yoga, art displays, and more. These all serve to create a sense of community through formal events, and informal happenings such as block parties or progressive dinners. (My friend who lives in Laureate Park has attended many of the latter.)

Example 3: some anti-branding

I realize that talking about branding in the public sector in general is difficult because it seems like a trick to divert resources toward something ephemeral and without much substance. Whenever I present about place branding at public administration conferences, I still get weird looks. The bottom line is branding and marketing are now integral parts of public sector life – but that does not mean the practices are without critique. I share stories here from a community develop-ment expert against some aspects of neighborhood branding, along with views from an economic development expert about the same.

The first person is an executive in a Washington, DC, community organizing group. He is the same person I quoted at length earlier about the redevelopment in the District's Chinatown. When I asked him why a neighborhood would want a brand identity, he said, "Don't brand your neighborhood." He said this is particu-larly the case for lower-income neighborhoods or historically minority enclaves that risk losing their authentic identities when someone comes in to derive, develop, and communicate a brand. Oftentimes, the conversation turns toward practices that spur gentrification and push people out.

He gave an example of a redevelopment project in nearby Silver Spring, Mary-land. He said there was a massive downtown redevelopment about 15 years ago that wiped out a large plaza where primarily black and immigrant families gath-ered. Instead of focusing on development ideas that would have benefitted the community, "all the development is chain stores and chain restaurants." A more equitable solution would have been really investing in the infrastructure and train-ing programs to give jobs to the local community, making the development sus-tainable as people developed skills they could pass along into perhaps their own businesses one day.

As a counter to this kind of market-based branding and redevelopment approach, he told me about the Dudley Street neighborhood in Boston. He said: "Instead of doing a technocratic, market-based approach, community organizers . . . got emi-nent domain authority to force land owners to sell vacant property. They got the state to assign eminent domain authority to this nonprofit so they became a major land owner." Most of that land and the buildings are in a community land trust, keeping some of the housing affordable and focusing on community-led redevel-opment concerns. "If you look across the street from where Dudley's imprint is, those homes have gentrified so the people who live there have been pushed out and displaced, but Dudley Street still retains the same sort of feel."

An economic development professional also in Washington, DC, also ques-tioned some of the motives for wanting a neighborhood brand. He explained:

Every single neighborhood already has a brand and identity. All a brand is is how people feel about something. So if you have a neighborhood with a bunch of rich people with big houses, the identity of the neighborhood is a bunch of rich people with big houses. There's other things that can paint that brand identity. If anybody says that neighborhoods don't or shouldn't have brand identity, you're fighting against the air. That's a misnomer to begin with. Sometimes those identities can be incorrect. Sometimes those identities can be harmful.

While he is not against neighborhood branding, he does want people to question why they would want or need a brand image apart from what already exists in people's minds. He said there are many lenses through which to view neighborhood branding – economic, social, real estate, for instance. Each person in those buckets has a different view of the goals, and not asking them to participate in meaningful ways in a rebranding process sees the results coming up short. "It comes back to your goal of why do you want to create a brand in general?"

He told me about a project to rebrand the Dupont Circle neighborhood in Washington, DC, which used to be known as the cool, hip neighborhood but has since "lost its mojo." He called city and neighborhood branding projects a "prisoner's dilemma" because "do you want cities competing versus cities for resources?" (A prisoner's dilemma explains why parties might compete even when the outcomes of cooperation would be in their best interests.) In DC, "every single neighborhood is spending a lot of money marketing" so they are competing against one another for the latest hip and cool label. "People aren't necessarily in general going to shop more so you're becoming more competitive and fighting against each other, and folks who don't have the resources to do that are hurt the most."

That is what happened to Dupont Circle, once a mainstay neighborhood that lost residents to competition with other areas. "It was an emerging neighborhood. It had a really vibrant gay scene, and now it's not the gay neighborhood anymore. It's kind of more stale and old and mainstream now." So a new business improvement district (BID) developed in Dupont Circle meant to turn around some of this brand image perception. "One of the fundamental things that the BID will do is marketing, branding, and place making and those sorts of things. That's really the right mechanism to have to make actual changes within the neighborhood that in reality strengthen their brand."

The process is still ongoing and will ideally involve community stakeholders. He noted that being a stable neighborhood is not a bad thing, but today's younger consumers are looking for the latest and greatest, so neighborhoods feel the pressure to keep up. This will only increase the prisoner's dilemma.

Concluding remarks

Doing the research for this book allowed me to talk to many people about their neighborhoods. I heard stories of love, stories of hope, and stories of change. Some people really never saw themselves moving, while others knew they were only on a temporary stop. I talked to people in all stages of life – young millennials, parents with children at home, empty nesters. I talked to people of varied socio-economic status. What remained constant across all the demographics was a desire for a place to build connections, live a vibrant life however they defined that, and feel safe calling home.

I hope this book can open up a conversation about the relationship between people and their neighborhoods that goes beyond tropes of the creative class and suburbia. There is no one right answer to the neighborhood identity question. Neighborhoods are so personal to people, and they tend to bristle when neighborhood changes happen that also shift their personal identities (Tyson, 2014). The research opens up a new area for public administration scholars to look beyond neighborhood governance issues and toward co-production ones with this micro-level unit. For scholars in tourism and management, we can now delve deeper into the branding processes taking place in neighborhoods, using those as in-depth case studies to see how people can shape an identity from the ground up – or how they respond when one is foisted on them.

I leave you with some quotes from people telling me what they like most about their neighborhoods. Hopefully you might see something here that resonates with you.

> *To me an ideal neighborhood is a couple of things. Number 1, location is big for me. I can get to work in five minutes, or I can get to the grocery store in five minutes, and I can get to pretty much everywhere in town in 15–20 minutes so that's important. It's people walking on the street, people playing with their kids outside where there's activity, where there's a sense of life. When I think of non-neighborhood, everybody stays inside all the time. You don't know your neighbors or you never see them. I really think of a neighborhood sort of like a community within your own community. What ends up happening is I go to the same grocery store five nights of the week; we go to the same restaurants that are in our neighborhood. You end up living in the area.*
> *– Convention and Visitors Bureau executive,*
> *Lafayette, Louisiana*

> *Our downtown is maybe relative to what you're used to, but we think it's pretty vibrant right now. When I was growing up, unless there was an Indians or Cavaliers or Browns game, you got the hell out of downtown. There was nobody there. There was nobody living downtown. There was a perception it was dangerous. I talk to people downtown, and it's not dangerous at all. . . . They've done, I think, a tremendous job downtown, but they have a long way to go.*
> *– Cleveland, Ohio, resident*

> *It's (branding) a slow, slow, slow process. It just is. Plastering a logo on everything is not going to make an immediate difference. I think as a city we have to work together to play up those areas and find what the good things*

are. In terms of the neighborhoods, most of the people that are involved in the neighborhood associations process already love their neighborhood; that's why they're involved.

— City official, Iowa City, Iowa

I actually think that our neighborhood is ideal for many reasons. In fact, from what I read it's what everybody is looking for. They like a trail; they like being near the water, parks, schools, transportation. We have all of that. The only thing I think would be really cool that we don't have is a little coffee shop or something.

— Millers Bay, Oshkosh, Wisconsin, resident

I like a neighborhood that's quiet, but I like a downtown area. I like to walk. I don't necessarily have to like to take my car out everywhere I go. I want good neighbors. That doesn't mean that they're all my friends. I want people who get involved when they're needed and mind their own business when they're not. I want to have choices in the neighborhood. I don't want to go to a mall for everything I buy. I like to shop locally even if I pay a little more for it. We have a lot of friends who are local merchants in the area. We support them. We support their stores. I like a town that has different aspects.

— Mont Clair, New Jersey, resident

References

Agranoff, R., and McGuire, M. (2001). Big questions in public network management research. *Journal of Public Administration Research and Theory*, *11*(3), 295–326.

Aitken, R., and Campelo, A. (2011). The four Rs of place branding. *Journal of Marketing Management*, *27*(9–10), 913–933.

Ananat, E. O. (2011). The wrong side(s) of the tracks: The causal effect of racial segregation on urban poverty and inequality. *American Economic Journal: Applied Economics*, *3*(2), 34–66.

Andehn, M., Kazeminia, A., Lucarelli, A., and Sevin, E. (2014). User-generated place brand equity on Twitter: The dynamics of brand associations in social media. *Place Branding and Public Diplomacy*, *10*(2), 132–144.

Andersson, I. (2016). 'Green cities' going greener? Local environmental policy-making and place branding in the 'Greenest City in Europe.' *European Planning Studies*, *24*(6), 1197–1215.

Anholt, S. (2010). Definitions of place branding – Working towards a resolution. *Place Branding and Public Diplomacy*, *6*(1), 1–10.

Anholt, S. (2005). Some important distinctions in place branding. *Place Branding*, *1*(2), 116–121.

Austin, D. M., Furr, L. A., and Spine, M. (2002). The effects of neighborhood conditions on perceptions of safety. *Journal of Criminal Justice*, *30*(5), 417–427.

Avraham, E., and Ketter, E. (2008). Will we be safe there? Analyzing strategies for altering unsafe place images. *Place Branding and Public Diplomacy*, *4*(3), 196–204.

Barnes, K., Waitt, G., Gill, N., and Gibson, C. (2006). Community and nostalgia in urban revitalization: A critique of urban village and creative class strategies as remedies for social 'problems.' *Australian Geographer*, *37*(3), 335–354.

Beck, H. (2009). Linking the quality of public spaces to quality of life. *Journal of Place Management & Development*, *2*(3), 240–248.

Bell, C. (2008). 100% Pure New Zealand: Branding for back-packers. *Journal of Vacation Marketing*, *14*(4), 345–355.

Bennetts, H., et al. (2017). Feeling safe and comfortable in the urban environment. *Journal of Urbanism: International Research on Placemaking and Urban Sustainability*, *10*(4), 401–421.

Bereitschaft, B. (2017). Equity in neighbourhood walkability? A comparative analysis of three large U.S. cities. *Local Environment*, *22*(7), 859–879.

Bilger, M., and Carrieri, V. (2011). Health in the cities: When the neighborhood matters more than income. *Journal of Health Economics*, *32*(1), 1–11.

Bonaiuto, M., et al. (1999). Multidimensional perception of residential environment quality and neighbourhood attachment in the urban environment. *Journal of Environmental Psychology*, *19*, 331–352.

Boyd, M. (2000). Reconstructing Bronzeville: Racial nostalgia and neighborhood redevelopment. *Journal of Urban Affairs*, *22*(2), 107–122.

Braun, E., Kavaratzis, M., and Zenker, S. (2013). My city – My brand: The role of residents in place branding. *Journal of Place Management and Development*, *6*(1), 18–28.

Brown, B., et al. (2003). Place attachment in a revitalizing neighborhood: Individual and block level of analysis. *Journal of Environmental Psychology*, *23*, 259–271.

Brown-Sarancino, J. (2004). Social preservationists and the quest for authentic community. *City & Community*, *3*(2), 135–156.

Brudney, J., and England, R. E. (1982). Urban policy making and subjective service evaluations: Are they compatible? *Public Administration Review*, *42*(2), 127–135.

Byrne, J. P. (2013). The rebirth of the neighborhood. *Fordham Urban Law Journal, 40*, 1595–1609.

Chan, C. (2017). Health-related elements in green space branding in Hong Kong. *Urban Forestry & Urban Greening, 21*, 192–202.

Chan, C., and Marafa, L. M. (2016). The green branding of Hong Kong: Visitors' and residents' perceptions. *Journal of Place Management and Development*, *9*(3), 289–312.

Chaskin, R. J., and Abunimah, A. (1999). A view from the city: Local government perspectives on neighborhood-based governance in community-building initiatives. *Journal of Urban Affairs*, *21*(1), 57–78.

Chaskin, R. J., and Garg, S. (1997). The issue of governance in neighborhood-based initiatives. *Urban Affairs Review, 32*(5), 631–661.

Choguill, C. L. (2008). Developing sustainable neighbourhoods. *Habitat International*, *32*, 41–48.

City of Orlando. (2019). *GreenUp Orlando*. Retrieved from www.cityoforlando.net/parks/green/

City of Painesville. (2019). *Painesville launches new brand celebrating city's history and diversity*. Retrieved from www.painesville.com/index.asp?SEC=7A3C2F08-E80F-4721-946A-B81E2640D593&DE=80C14B1B-414D-4BF9-B080-91F50A9B7971&Type=B_PR

Coaffee, J., and Rogers, P. (2008). Reputational risk and resiliency: The branding of security in place-making. *Place Branding and Public Diplomacy*, *4*(3), 205–217.

Coates, T. (2009). *Some explanation is due*. Retrieved from www.theatlantic.com/entertainment/archive/2009/04/some-explanation-is-due/16423/

Cochrun, S. E. (1994). Understanding and enhancing neighborhood sense of community. *Journal of Planning Literature*, *9*(1), 92–99.

Comstock, N., et al. (2010). Neighborhood attachment and its correlates: Exploring neighborhood conditions, collective efficacy, and gardening. *Journal of Environmental Psychology*, *30*, 435–442.

Crompton, J. L. (2001). The impact of parks on property values: A review of the empirical evidence. *Journal of Leisure Research, 33*(1), 1–31.

de Azvedo, A. J. A., et al. (2013). "Are you happy here?": The relationship between quality of life and place attachment. *Journal of Place Management and Development*, *6*(2), 102–119.

Deener, A. (2007). Commerce as the structure and symbol of neighborhood life: Reshaping the meaning of community in Venice, California. *City & Community*, *6*(4), 291–314.

Dekker, K., and Varady, D. P. (2011). A comparison of Dutch and US public housing regeneration planning? The similarity grows? *Urban Research & Practice*, *4*(2), 123–152.

Diez Roux, A. V. (2001). Investigating neighborhood and area effects on health. *American Journal of Public Health, 91*(11), 1783–1789.

Domonske, C. (2014). *Segregated from its history, how 'ghetto' lost its meaning*. Retrieved from www.npr.org/sections/codeswitch/2014/04/27/306829915/segregated-from-its-history-how-ghetto-lost-its-meaning

Eimermann, M. (2015). Promoting Swedish countryside in the Netherlands: International rural place marketing to attract new residents. *European Urban and Regional Studies*, *22*(4), 398–415.

Ellis, C. (2002). The new urbanism: Critiques and rebuttals. *Journal of Urban Design*, *7*(3), 261–291.

Eshuis, J., Braun, E., and Klijn, E. (2013). Place marketing as governance strategy: An assessment of obstacles in place marketing and their effects on attracting target groups. *Public Administration Review*, *73*(3), 507–516.

Eshuis, J., and Edelenbos, J. (2009). Branding in urban regeneration. *Journal of Urban Regeneration and Renewal*, *2*(3), 272–282.

Eshuis, J., and Klijn, E. (2012). *Branding in governance and public management*. London: Taylor & Francis.

Evans-Cowley, J. S. (2010). Planning in the age of Facebook: The role of social networking in planning processes. *GeoJournal*, *75*(5), 407–420.

Farris, J., and Kendrick, A. (2011). *Neighborhood brands*. Dallas, TX: BubbleLife Media.

Fay, D. L., and Zavattaro, S. M. (2016). Branding and isomorphism: The case of higher education. *Public Administration Review*, *76*(5), 805–815.

Florida, R. (2002). *The rise of the creative class*. New York: Basic Books.

Foucault, M. (1977). *Discipline and punish*. New York, NY: Vintage.

Frumkin, H. (2003). Healthy places: Exploring the evidence. *American Journal of Public Health, 93*(9), 1451–1456.

Gascon, M., et al. (2015). Mental health benefits of long-term exposure to residential green and blue spaces: A systemic review. *International Journal of Environmental Research and Public Health*, *12*, 4354–4379.

Glassman, C. (2019). *'We are not Tribeca!': In Belfast, battle over a neighborhood rebranding*. Retrieved from www.tribecatrib.com/content/we-are-not-tribeca-belfast-battle-over-neighborhood-rebranding

Goetze, R., and Colton, K. W. (1980). The dynamics of neighborhoods: A fresh approach to understanding housing and neighborhood change. *Journal of the American Planning Association*, *46*(2), 184–194.

Gotham, K. F. (2002). Marketing Mardi Gras: Commodification, spectacle, and the political economy of tourism in New Orleans. *Urban Studies*, *39*(10), 1735–1756.

Gringlas, S. (2017). *Old confronts new in a gentrifying D.C. neighborhood*. Retrieved from www.npr.org/2017/01/16/505606317/d-c-s-gentrifying-neighborhoods-a-careful-mix-of-newcomers-and-old-timers

Gulsrud, N. M., et al. (2013). Green space branding in Denmark in the era of neoliberal governance. *Urban Forestry & Urban Greening*, *12*, 330–337.

Haaland, C., and van den Bosch, C. K. (2015). Challenges and strategies for urban green-space planning in cities undergoing densification: A review. *Urban Forestry & Urban Greening*, *14*(4), 760–771.

Hampton, K., and Wellman, B. (2003). Neighboring in Netville: How the internet supports community and social capital in a wired suburb. *City & Community*, *2*(4), 277–311.

Hanna, S., and Rowley, J. (2008). An analysis of terminology used in place branding. *Place Branding and Public Diplomacy*, *4*(1), 61–75.

Harding, D. L. (2003). Counterfactual models of neighborhood effects: The effect of neighborhood poverty on dropping out and teenage pregnancy. *American Journal of Sociology*, *109*(3), 676–719.

Hartmann, P., Ibanez, V. A., and Sainz, F. J. F. (2005). Green branding effects on attitude: Functional versus emotional positioning strategies. *Marketing Intelligence & Planning*, *23*(1), 9–29.

Havlena, W. J., and Holak, S. L. (1991). The 'good old days': Observations on nostalgia and its role in consumer behaviour. *Advances in Consumer Research*, *18*(1), 323–339.

Hidalgo, M. C., and Hernandez, B. (2001). Place attachment: Conceptual and empirical questions. *Journal of Environmental Psychology*, *21*(3), 273–281.

Hoffer, A. (2015). *Where we live: NoMa, the wrong side of the tracks no more*. Retrieved from www.washingtonpost.com/realestate/where-we-live-noma-the-wrong-side-of-the-tracks-no-more/2015/02/05/b85daf88-a0d5–11e4–903f-9f2faf7cd9fe_story.html?noredirect=on&utm_term=.64dc140afa9e

Hunt, L., and Johns, N. (2013). Image, place and nostalgia in hospitality branding and marketing. *Worldwide Hospitality and Tourism Themes*, *5*(1), 14–26.

Hutta, J. S. (2009). Geographies of *Geborgenheit*: Beyond feelings of safety and the fear of crime. *Environment and Planning D: Society and Space*, *27*, 251–273.

Jackson, L. E. (2003). The relationship of urban design to human health and condition. *Landscape Urban Plann*, *64*(4), 191–200.

Jacobs, J. (1961). *The death and life of great American cities*. New York, NY: Vintage Books.

Johansson, O., and Cornebise, M. (2010). Place branding goes to the neighbourhood: The case of pseudo-Swedish Andersonville. *Geografiska Annaler*, *92*(3), 187–204.

Johnson, B. J., and Halegoua, G. R. (2014). Potential and challenges for social media in the neighborhood context. *Journal of Urban Technology*, *21*(4), 51–75.

Kaczynski, A. T., Potwarka, L. R., and Saelens, B. E. (2008). Association of park size, distance, and features within physical activity in neighborhood parks. *American Journal of Public Health*, *98*(8), 1451–1456.

Kallus, R., and Law-Yone, H. (2000). What is a neighbourhood? The structure and function of an idea. *Environment and Planning B: Planning and Design*, *27*, 815–826.

Karsten, L. (2014). From yuppies to yupps: Family gentrifiers consuming spaces and re-inventing cities. *Tijdschrift vor Economische en Sociale Geografie*, *105*(2), 175–188.

Kasinitz, P., and Hillyard, D. (1995). The old-timers' tale: The politics of nostalgia on the waterfront. *Journal of Contemporary Ethnography*, *24*(2), 139–164.

Kavaratzis, M., and Hatch, M. J. (2013). The dynamics of place brands: An identity-based approach to place branding theory. *Marketing Theory*, *13*(1), 69–86.

Keatinge, B., and Martin, D. G. (2015). A 'Bedford Falls' kind of place: Neigbourhood branding and commercial revitalization in processes of gentrification in Toronto, Ontario. *Urban Studies*, 867–883.

Kelling, G. L., and Wilson, J. Q. (1982). *Broken windows*. Retrieved from www.theatlantic.com/magazine/archive/1982/03/broken-windows/304465/

Kelsh, J. (2015). *Neighborhood branding and marketing*. Retrieved from https://www.neighborworks.org/Documents/Community_Docs/Revitalization_Docs/StableCommunities_Docs/Defining-the-New-Brand.aspx

Kemperman, A., and Timmermans, H. (2014). Green spaces in the direct living environment and social contacts of the aging population. *Landscape Urban Planning*, *129*, 44–54.

Kenny, J., and Zimmerman, J. (2004). Constructing the "genuine American city": Neotraditionalism, New Urbanism, and neo-liberalism in the remaking of downtown Milwaukee. *Cultural Geographies, 11*(1), 74–98.

Klijn, E., Eshuis, J., and Braun, E. (2012). The influence of stakeholder involvement in the effectiveness of place branding. *Public Management Review, 14*(4), 499–519.

Koskela, H., and Pain, R. (2000). Revisiting fear and place: Women's fear of attack and the built environment. *Geoforum, 31*(2), 269–280.os

Kotler, P., Haider, D. H., and Rein, I. (1993). *Marketing places: Attracting investment, industry and tourism to cities, States and Nations.* New York, NY: The Free Press.

Lee, A. C. K., and Maheswaran, R. (2011). The health benefits of urban green spaces: A review of the evidence. *Journal of Public Health, 33*(2), 212–222.

Lichrou, M., O'Malley, L., and Patterson, M. (2010). Narratives of a tourism destination: Local particularities and their implications for place marketing and branding. *Place Branding and Public Diplomacy*, 6(2), 134–144.

Lo, A. Y. H., and Jim, C. Y. (2012). Citizen attitude and expectations towards greenspace provision in compact urban milieu. *Land Use Policy, 29*, 577–586.

Logan, J. R., and Rabrenovic, G. (1990). Neighborhood associations: Their issues, their allies, and their opponents. *Urban Affairs Quarterly, 26*(1), 68–94.

Lovrich, N. P., and Taylor, G. T. (1976). Neighborhood evaluation of local government services: A citizen survey approach. *Urban Affairs Quarterly, 12*(2), 197–222.

Maas, J., et al. (2009a). Morbidity is related to a green living environment. *Journal of Epidemiology and Community Health, 63*(12), 967–973.

Maas, J., et al. (2009b). Social contacts as a possible mechanism behind the relation between green space and health. *Health & Place, 15*(2), 586–595.

Macmillan, R., et al. (2000). Experiencing the streets: Harassment and perceptions of safety among women. *Journal of Research in Crime and Delinquency, 37*(3), 306–322.

Majumdar, S., Deng, J., Zhang, Y., and Pierskalla, C. (2012). Using contingent valuation to estimate the willingness of tourists to pay for urban forests: A study in Savannah, Georgia. *Urban Forestry and Urban Greening, 10*(4), 275–280.

Marston, S. A. (1988). Neighborhood and politics: Irish ethnicity in nineteenth century lowell, Massachusetts. *Annals of the Association of American Geographers, 78*(3), 414–432.

Masden, C., et al. (2014). *Tensions in scaling-up community social media: A multineighborhood study of Nextdoor.* Proceedings of the 32nd annual ACM conference on Human factors in computing systems, pp. 239–248.

Masuda, J. R., and Bookman, S. (2018). Neighbourhood branding and the right to the city. *Progress in Human Geography, 42*(2), 165–182.

McCann, E. (2013). Policy boosterism, policy mobilities, and the extrospective city. *Urban Geography, 34*(1), 5–29.

McClinchey, K. A. (2008). Urban ethnic festivals, neighborhoods, and the multiple realities of marketing place. *Journal of Travel and Tourism Marketing, 25*(3–4), 251–264.

McMillan, D. W., and Chavis, D. M. (1986). Sense of community: A definition and theory. *Journal of Community Psychology, 14*, 6–23.

Melvin, P. M. (1985). Changing contexts: Neighborhood definition and urban organization. *American Quarterly, 37*(3), 357–367.

Mitchell, R., and Popham, F. (2007). Greenspace, urbanity and health: Relationships in England. *Journal of Epidemiology and Community Health, 61*, 681–683.

Morrill, R. L. (1965). The Negro ghetto: Problems and alternatives. *Geographical Review, 55*(3), 339–361.

Mumford, L. (1954). The neighborhood and the neighborhood unit. *The Town Planning Review*, *24*(4), 256–270.

National Geographic Magazine. (2018). *Neighborhood*. Retrieved from www.nationalgeo graphic.org/encyclopedia/neighborhood/

Ntounis, N., and Kavartzis, M. (2017). Re-branding the High Street: The place branding process and reflections from three UK towns. *Journal of Place Management and Development*, *10*(4), 392–403.

Oberai, H., and Anand, I. M. (2018). Unconscious bias: Thinking without thinking. *Human Resource Management International Digest*, *26*(6), 14–17.

Ocejo, R. E. (2011). The early gentrifier: Weaving a nostalgia narrative on the Lower East Side. *City & Community*, *10*(3), 285–310.

Osborne, D. E., and Gaebler, T. (1993). *Reinventing government: How the entrepreneurial spirit is transforming the public sector*. New York, NY: Penguin Random House.

Pain, R. H. (2001). Gender, race, age and fear in the city. *Urban Studies*, *38*(5–6), 899–913.

Pain, R. H. (1997). Social geographies' of women's fear of crime. *Transactions of the Institute of British Geographers*, *22*(2), 231–244.

Pais, J., Batson, C. D., and Monnat, S. M. (2014). Neighborhood reputation and resident sentiment in the wake of the Las Vegas foreclosure crisis. *Sociological Perspectives*, *57*(3), 343–363.

Peck, J. (2005). Struggling with the creative class. *International Journal of Urban and Regional Research*, *29*(4), 740–770.

Perkins, D. D., and Long, D. A. (2002). Neighborhood sense of community and social capital: A multi-level analysis. In A. Fisher, C. Sonn, and B. Bishop (Eds.), *Psychological sense of community: Research, applications, and implications* (pp. 291–318). New York, NY: Plenum.

Permentier M., van Ham, M., and Bolt, G. (2009). Neighbourhood reputation and the intention to leave the neighbourhood. *Environment and Planning A*, *41*, 2162–2180.

Pickering, M., and Keightley, E. (2006). The modalities of nostalgia. *Current Sociology*, *54*(6), 919–941.

Pinjari, A. R., et al. (2008). Joint model of choice of residential neighborhood and bicycle ownership. *Transportation Research Record: Journal of the Transportation Research Board*, *2082*, 17–26.

Powell, L. M. et al (2006). Availability of physical activity-related facilities and neighborhood demographic and socioeconomic characteristics: A national study. *American Journal of Public Health*, *96*(9), 1676–1680.

Pretty, J., Peacock, J., Sellens, M., and Griffin, M. (2005). The mental and physical health outcomes of green exercise. *International Journal of Environmental Health Research*, *15*(5), 319–337.

Quillian, L. (2008). Does unconscious racism exist? *Social Psychology Quarterly*, *71*(1), 6–11.

Quillian, L. (2003). How long are exposures to poor neighborhoods? The long-term dynamics of entry and exit from poor neighborhoods. *Population Research and Policy Review*, *22*, 221–249.

Quillian, L., and Prager, D. (2001). Black neighbors, higher crime? The role of racial stereotypes in evaluations of neighborhood crime. *American Journal of Sociology*, *107*(3), 717–767.

Renn, A. (2019). *Cities: Don't fall for the branding trap*. Retrieved from www.citylab.com/ perspective/2019/02/city-branding-traps-marketing-campaigns/582496/

Rich, M. A., and Tsistos, W. (2016). Avoiding the "SoHo effect" in Baltimore: Neighborhood revitalization and arts and entertainment districts. *International Journal of Urban and Regional Research, 40*(4), 736–756.

Robinson, D., and Wilkinson, D. (1995). Sense of community in a remote mining town: Validating a neighborhood cohesion scale. *American Journal of Community Psychology, 23*(1), 137–148.

Roe, J. J., Aspinall, P. A., and Ward Thompson, C. (2017). Coping with stress in deprived urban neighborhoods: What is the role of green space according to life stage? *Frontiers in Psychology.* https://doi.org/10.3389/fpsyg.2017.01760

Roe, J. J., et al. (2013). Green space and stress: Evidence from cortisol measures in deprived urban communities. *International Journal of Environmental Research in Public Health, 10*, 4086–4103.

Rogers, G. O., and Sukolratanametee, S. (2009). Neighborhood design and sense of community: Comparing suburban neighborhoods in Houston Texas. *Landscape and Urban Planning, 92*, 325–334.

Rothstein, R. (2017). *The color of law: A forgotten history of how our government segregated America.* New York, NY: Liveright Publishing.

Sampson, R. J., Raudenbush, S. W., and Earls, F. (1997). Neighborhoods and violent crime: A multilevel study of collective efficacy. *Science, 277*(5328), 918–924.

Sampson, R.J., and Sharkey, P. (2008). Neighborhood selection and the social reproduction of concentrated racial inequality. *Demography, 45*(1), 1–29.

Sandberg, L., and Ronnblom, M. (2015). 'I don't think we'll ever be finished with this': Fear and safety in policy and practice. *Urban Studies, 52*(14), 2664–2679.

Santala, V., et al. (2017). *Making sense of the city: Exploring the use of social media data for urban planning and place branding.* Paper presented at Anais Do I Workshop de Computacao Urbana, Brazil.

Sastry, N., and Pebley, A. R. (2010). Family and neighborhood sources of socioeconomic inequality in children's achievements. *Demography, 47*(3), 777–800.

Schilling, J., and Logan, J. (2008). Greening the Rust Belt: A green infrastructure model for right sizing America's shrinking cities. *Journal of the American Planning Association, 74*(4), 451–466.

Sevin, E. (2014). Understanding cities through city brands: City branding as a social and semantic network. *Cities, 38*, 47–56.

Sevin, E. (2011). Thinking about place branding: Ethics of concept. *Place Branding and Public Diplomacy, 7*(3), 155–164.

Shirlow, P., and Pain, R. H. (2003). The geographies and politics of fear. *Capital & Class, 27*(2), 15–26.

Smeltz, A., and Murray, A. (2019). *Pittsburgh council balks over six-figure branding proposal.* Retrieved from www.post-gazette.com/local/city/2018/12/08/Pittsburgh-council-balks-over-branding-proposal-marketing-city/stories/201812070110

Smith, C. A., and Smith, C. J. (1978). Locating natural neighbors in the urban community. *Area, 10*(2), 358–369.

Sneed, C. T., Runyan, R., Swinney, J. L., and Lim, H. (2011). Brand, business mix, sense-of-place: Do they matter downtown? *Journal of Place Management and Development, 4*(2), 121–134.

Strange, J. H. (1972). Citizen participation in community action and model cities programs. *Public Administration Review, 32*(Special Issue), 655–669.

Talen, E. (1999). Sense of community and neighbourhood form: An assessment of the social doctrine of new urbanism. *Urban Studies, 36*(8), 1361–1379.

Tandogan, O., and Ilhan, B. S. (2016). Fear of crime in public spaces: From the viewpoint of women living in cities. *Procedia Engineering, 161,* 2011–2018.

Tannenbaum, J. (1948). The neighborhood: A socio-psychological analysis. *Land Economics, 24*(4), 358–369.

Taylor, R. B., and Covington, J. (1993). Community structural change and fear of crime. *Social Problems, 40*(3), 374–397.

Taylor, R. B., Shumaker, S. A., and Gottfredson, S. D. (1985). Neighborhood-level links between physical features and local sentiments: Deterioration, fear of crime, and confidence. *Journal of Architectural and Planning Research, 2*(4), 261–275.

Turner, B. (1987). A note on nostalgia. *Theory Culture and Society, 4,* 147–156.

Tyson, C. J. (2014). Municipal identity as property. *Penn State Law Review, 118*(3), 647–696.

Ujang, N. (2012). Place attachment and continuity of urban place identity. *Procedia – Social and Behavioral Sciences, 49,* 156–167.

Ulldemolins, J. R. (2014). Culture and authenticity in urban regeneration processes: Place branding in central Barcelona. *Urban Studies, 51*(14), 3026–3045.

Unger, D. G., and Wandersman, A. (1985). The importance of neighbors: The social, cognitive, and affective components of neighboring. *American Journal of Community Psychology, 13*(2), 139–169.

Unger, D. G., and Wandersman, A. (1983). Neighboring and its role in block organizations: An exploratory report. *American Journal of Community Psychology, 11*(3), 291–300.

U.S. Census Bureau. (2012). *Geographic terms and concepts – Census tracts.* Retrieved from www.census.gov/geo/reference/gtc/gtc_ct.html

U.S. Department of Housing and Urban Development. (n.d.). *The Model Cities Program: Q&A.* Retrieved from https://ia800502.us.archive.org/13/items/modelcitiesprogr00unit/modelcitiesprogr00unit.pdf

Valentine, G. (2007). Theorizing and researching intersectionality: A challenge for feminist geography. *The Professional Geographer, 59*(1), 10–21.

van Ryzin, G. G. (2004). Expectations, performance, and citizen satisfaction with urban services. *Journal of Policy Analysis and Management, 23*(3), 433–448.

Varady, D. P. (1986). Neighborhood confidence: A critical factor in neighborhood revitalization?. *Environment and Behavior, 18*(4), 480–501.

Watson, S., and Wells, K. (2005). Spaces of nostalgia: The hollowing out of a London market. *Social & Cultural Geography, 6*(1), 17–30.

Wherry, F. W. (2011). *The Philadelphia barrio: The arts, branding, and neighborhood transformation.* Chicago, IL: University of Chicago Press.

Wilkinson, A. (2018). *To Trump fans, #MAGA is more than a slogan. It's an aesthetic.* Retrieved from https://www.vox.com/culture/2018/8/8/17376824/trump-fan-art-maga-dinesh-dsouza-jon-mcnaughton

Willis, A. (2019). *Spring Hill considers signing $66,000 contract with marketing organization for "place branding".* Retrieved from https://springhillhomepage.com/spring-hill-considers-signing-66000-contract-with-marketing-organization-for-place-branding/

Wilson, W. J. (1991). Studying inner-city social dislocations: The challenge of public agenda research. *American Sociological Review, 56*(1), 1–14.

Wood, L., et al. (2008). The anatomy of the safe and social suburb: An exploratory study of the built environment, social capital and residents' perceptions of safety. *Health & Place, 14,* 15–31.

Wood, L., Frank, L. D., and Giles-Corti, B. (2010). Sense of community and its relationship with walking and neighborhood design. *Social Science & Medicine, 70*(9), 1381–1390.

Woodard, C. (2016). *How Cincinnati salvaged the nation's most dangerous neighborhood.* Retrieved from www.politico.com/magazine/story/2016/06/what-works-cincinnati-ohio-over-the-rhine-crime-neighborhood-turnaround-city-urban-revitalization-213969

Woolever, C. (1992). A contextual approach to neighbourhood attachment. *Urban Studies, 29*(1), 99–116.

Yang, Y. (2008). A tale of two cities: Physical form and neighborhood satisfaction in metropolitan Portland and Charlotte. *Journal of the American Planning Association, 74*(3), 307–323.

Yates, D. (1972). Neighborhood government. *Policy Sciences, 3,* 209–217.

Yoo, S. (2018). *People consider re-branding of Minneapolis' oldest neighborhood.* Retrieved from www.kare11.com/article/news/people-consider-re-branding-of-minneapolis-oldest-neighborhood/89-572870072

Zavattaro, S. M. (2013). *Cities for sale.* Albany, NY: SUNY Press.

Zavattaro, S. M., and Fay, D. L. (2019). Brand USA: A natural quasi-experiment evaluating the success of a national marketing campaign. *Tourism Management, 70,* 42–48.

Zenker, S. (2011). How to catch a city? The concept and measurement of place brands. *Journal of Place Management and Development, 4*(1), 40–52.

Zenker, S., and Martin, N. (2011). Measuring success in branding and marketing. *Place Branding and Public Diplomacy, 7*(1), 32–41.

Zhang, Y., van Dijk, T., Tang, J., and Berg, A. E. (2015). Green space attachment and health: A comparative study in two urban neighborhoods. *International Journal of Environmental Research and Public Health, 12*(11), 14342–14363.

Zheng, S. (2017). *The rise and fall of D.C.'s Chinatown.* Retrieved from www.historians.org/publications-and-directories/perspectives-on-history/december-2017/the-rise-and-fall-of-dcs-chinatown

Zukin, S. (1996). *The culture of cities.* Oxford: Blackwell.

Zukin, S. (1987). Gentrification: Culture and capital in the urban core. *Annual Review of Sociology, 13,* 129–147.

Index